Vertebrate Embryology

A LABORATORY MANUAL

Vertebrate Embryology

A LABORATORY MANUAL
Third Edition

by Richard M. Eakin

UNIVERSITY OF CALIFORNIA PRESS
Berkeley and Los Angeles

UNIVERSITY OF CALIFORNIA PRESS
BERKELEY AND LOS ANGELES, CALIFORNIA

UNIVERSITY OF CALIFORNIA PRESS, LTD.
LONDON, ENGLAND

THIRD EDITION, REVISED, 1978

© 1947, 1971, 1978 by Richard M. Eakin
Library of Congress Catalog Card No.: 77–88420

ISBN 0–520–03593–3
Printed in the United States of America

PREFACE

The traveler exploring a foreign city for the first time uses a guidebook with a map to lead him to places of scenic, cultural, and historic note and with commentary to enlarge his appreciation of their significance. The architecture of a cathedral is explained, a monument elucidated, a masterpiece interpreted, the technique of the artist expounded, the costume of the people traced to ancient times. Similarly the student of vertebrate embryology needs a guide with a map of the embryo to lead him to structures of importance and commentary to enhance his understanding of their relationship to the plan of the adult body and to explain the unseen dynamic processes which shape the course of development.

This laboratory manual was written originally for classes in vertebrate embryology at the University of California. An increasing number of other institutions adopted it, and I was advised that it would be helpful to still others if a section on the frog larva were added. This I have done with the assistance of Mrs. Emily E. Reid, the scientific illustrator of the Department of Zoölogy at Berkeley. I have taken the opportunity to make other additions and some changes in the manual.

The preparation of the guide to the frog larva has been exciting, and I am impressed anew with the beauty of amphibian embryos and their instructive value in developmental anatomy and histogenesis. I commend frequent use of high power magnification in the study of differentiating tissues and organs after they have been identified and their gross relationships determined.

The revision of the manual unexpectedly required several weeks of a sabbatical leave supported in part by the United States Public Health Service. It is proper, therefore, to acknowledge my indebtedness to the Service for its grant-in-aid. Considering the relevance and importance of embryology to the health services of the nation I hope that the National Institutes of Health will be pleased to have contributed to the improvement of this teaching aid and to an extension of its usefulness. I have not forgotten those who assisted in the preparation of earlier editions, particularly Mrs. Lois C. Stone and Mr. John S. Gildersleeve, and students, assistants, and colleagues who made helpful suggestions.

RICHARD M. EAKIN

Berkeley, California
January, 1964

ADDENDUM

The need for a third printing of the manual gives me the opportunity to make some improvements: up-dating the discussion of a few topics such as induction and the origin of primordial germ cells, the enlargement of the text on the dissection of the fetal pig, and the addition of many new figures. I am exceedingly grateful to Emily Reid, our departmental artist, who transformed into text figures the sketches made by myself and by Ellen Townes, a graduate student. I acknowledge the assistance of Miss Aileen Kuda of my laboratory and that of Mr. Joel F. Walters, Editor, University of California Press.

R. M. E.

Berkeley, California
January, 1971

ADDENDUM

A substantial revision of the manual had been planned by adding more figures, a glossary, and further discussion of principles. The schedule for publication, however, limits me to correcting errors, up-dating, and including a few paragraphs of new text without disturbing the pagination. Nor am I sure that additional figures would contribute to good pedagogy. I think that both students and instructors are too addicted to picture books. We prefer to consult a labelled figure or photograph of every section (or every fifth one), to identify a structure, instead of reading a brief paragraph, then tracing and discovering an answer ourselves. However, I believe that if we take the easy path, we make a sacrifice. As stated in the original preface to this book, identification of a structure is not the only goal of a laboratory exercise in embryology—perhaps, not even the most important one.

I continue to learn from students and assistants. To them I again acknowledge my indebtedness. My last associate, Alan Rapraeger, merits special mention; and I am grateful to my research grant from the United States Public Health Service for partial support of this revision.

R. M. E.

Berkeley, California
September, 1977

CONTENTS

List of Summaries and Discussions of General Topics

INTRODUCTION

A brief statement on the importance of embryology in relation to other zoölogical subsciences and on the values of a course in vertebrate development should be helpful to the student.

If one takes the point of view that an understanding of function is a primary objective in biology, then the subscience of physiology assumes a position, in relation to other disciplines or areas of knowledge, indicated in figure 1. We see that physiology, represented by the topmost block, is supported by a foundation of other units arranged into two groups: on the right a pyramid of biological studies and, on the left, one based upon the physical sciences and mathematics. Although the relationships here pictured are obvious, let us consider more closely the connections between physiology, the anatomical sciences, the study of evolution, and embryology.

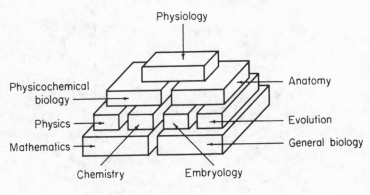

Fig. 1.

It is almost axiomatic that an understanding of function requires a knowledge of structure. Hence, the anatomical sciences—morphology, histology, cytology, and, one might add, pathology—are basic to physiology. Anatomy in turn rests upon history. In other words, one cannot fully understand or appreciate structure apart from its history, be it the structure of social institutions, government, mountain ranges, river systems, or the human body. The architecture of the body, as revealed by dissection and microscopical examination, has had a two-fold history: (1) the phylogenetic record of evolutionary changes on a macro time scale which have transformed primitive structural patterns into those of today, and (2) the ontogenetic record of developmental changes on a micro time scale which have converted simple embryonic patterns into those of the adult.

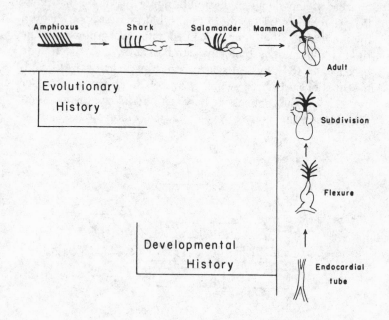

Fig. 2.

To illustrate this point, consider the two histories of the heart and aortic arches. The evolutionary history (see fig. 2, horizontal axis) is suggested by several kinds of vascular patterns found in chordates living today, four of which may be selected as points on the macro time scale. (1) Amphioxus. In this protochordate may be seen the ancestral prevertebrate pattern in which the pulsating branchial artery or ventral aorta is the forerunner of the heart and the many pairs of afferent arteries to the gill basket the precursors of the aortic arches. (2) Shark. The condition of heart and branchial vessels in this primitive vertebrate illustrates a second theoretical stage in the evolution of these structures. The heart shows clear subdivision into chambers—sinus venosus, atrium, ventricle, and conus—and a bending of the once straight tube. The branchial afferent arteries, five to seven pairs, are the old branchial vessels of Amphioxus, reduced in number. (3) Salamander. The urodele amphibian typefies a later stage in the evolutionary chornicle. The flexure of the heart is greater and the atrium has become subdivided into two auricles. The aortic arches are further reduced to four pairs. (4) Mammal. The history ends with the pattern in birds and mammals in which the ventricle as well as the atrium is subdivided and only three aortic arches—the third, fourth, and sixth—contribute to adult vessels: the carotids, the systemic arch, and the pulmonary arteries.

For the developmental history of the heart and aortic arches (fig. 2, vertical axis) observe the ontogeny of a bird or mammal. It will be similar in certain respects to the account just given. Again, four stages are selected for note. (1) The embryonic primordium of the heart is an uncoiled tube of endocardium. Its anterior bifurcated extension forms the ventral aorta from which develop, progessively from anterior to posterior, pairs of lateral branches, the aortic arches. (2) Flexure of the tube occurs and the future chambers begin to appear—sinus venosus, atrium, ventricle, and conus. The aortic arches become well-defined vessels carrying blood through the vis-

3

ceral arches to the dorsal aorta. As posterior pairs of arches are added, anterior ones fade; the total number is six pairs. (3) The heart is further coiled, the atrium now becomes divided into two chambers and subdivision of the ventricle is foreshadowed. Only three aortic arches persist. (4) The developmental history terminates with the adult picture of two auricles, two ventricles, and carotid, systemic, and pulmonary arteries derived from the third, fourth, and sixth aortic arches respectively.

A second example of the two kinds of history, usually mutually illuminating and often strikingly similar, is provided by the thyroid gland. The thyroid, an endocrine gland in modern vertebrates, probably evolved from an exocrine gland associated with the pharynx. Certain studies of the endostyles of Amphioxus and of the metamorphosing larval lamprey, ammocoetes, have suggested the origin of the thyroid from the food-concentrating apparatus of the protochordate. In the frilled shark, *Chlamydoselachus,* which is a very primitive elasmobranch, we find a thyroid gland still connected to the pharynx by an ancestral duct. In higher vertebrates, however, this connection of the adult thyroid with its evolutionary site of origin has disappeared. The phylogenic history, although here abbreviated, is at best fragmentary.

The other history, the embryology of the thyroid, is well documented by studies of the developmental origin of the gland and of the sequence of transformations leading to the structural pattern of the adult. The thyroid is traced from its origin as a ventral outpocketing of the pharynx, which in early stages of development is reminiscent of the gross appearance of the endostyle of Amphioxus (see fig. 43). The primordium elongates and bears for a time a thyroglossal duct to the pharynx; it subdivides and differentiates morphologically into lobes and histologically into thyroid follicles, stroma, and capsule. In the hands of the experimental embryologist the developmental origin of the thyroid can be explored in quite early, preprimordial stages of development in which the anlage of the thyroid shows no visible differentiation into even the rudiment of the gland. This is developmental history, the subject of our course.

The purposes and values of a course in embryology may be regarded as, first, specific and professional; second, general and cultural. Foremost, perhaps, in the first group of objectives is the acquisition of a large body of embryological information, the importance of which in relation to anatomy has been discussed above. This includes the broad outlines and the detailed facts about the development of the vertebrate body and certain extraembryonic structures. An aim closely related to the first is the understanding of developmental anomalies—from two-headed monstrosities to large and small irregularities in the position of organs or in the arrangement of blood vessels. Not only does embryology provide an explanation for the normal architectural patterns of the body, many of which would be complete riddles to the anatomist untrained in embryology, but for the abnormal ones as well. Also of importance for both advanced studies and professional work in either pure or applied biology is experience in mental imagery. Practice in this excellent technique of mastering and retaining detail is admirably afforded by a course in embryology, because the student is required to construct mental images of structures, visualized in three dimensions, whose forms are constantly changing during development. The method of rote memorization is hopelessly inadequate; the student must *see* these developmental patterns and should be able to translate his mental images into sketches and words. In constructing a three-dimensional figure the student will be working, for the most part, with two-dimensional views obtained from serial sections of embryos. By careful tracing, these sections are mentally fitted together to give the complete stereoscopic picture. This process is laborious and frequently discouraging to the student. In his eagerness to obtain the facts he overlooks the value of the method. Mental images formed slowly and in a stepwise manner, working from the known to the unknown, are more meaningful, more satisfying, and unquestionably more lasting.

The general values of a course in embryology are stated with greater uncertainty and difficulty. As in all of the natural sciences, so in embryology, the student is continually encouraged to give steadfast attention to accuracy, precision, independent and self-reliant work, and to neat and orderly arrangement of ideas, notes, and drawings; to receive guidance and stimulus from the opinions of others but to verify factual information and general impressions, so far as he may be able, by observations of his own. In addition, embryology presents the student with ever-changing and increasingly complex patterns of organization in which unity, order, harmonic relationship of many parts, and a faithfulness to hereditary blueprints are particularly striking.

I

GAMETOGENESIS
AND THE ESTROUS CYCLE

Although the embryo or new individual begins with the zygote, resulting from the union of the gametes, ovum and sperm, it is desirable to introduce our course with a review of gametogenesis, the process whereby the gametes are formed in the gonads, ovaries and testes, and of the estrous cycle (G. *oistros* = gadfly, hence, frenzy or desire), the cyclical changes occurring in the reproductive system of the female mammal.

A. Oögenesis in the Mammal

1. RAT OVARY: Examine microscopically the slide of rat ovary, first at intermediate and then at high magnification. Because histological and embryological preparations are expensive, care in handling slides can not be overemphasized. Most breakages are the result of the following mistakes. (1) Focusing the microscope upon a preparation by lowering the objective with the eye at the ocular. Instead, one should lower the barrel of the microscope while *watching from the side* to see that the objective does not touch the coverslip. Then with eye at the ocular raise the barrel until the object is brought into focus. If the optical plane is missed repeat the procedure. (2) Changing from intermediate to high power by rapid and careless rotation of the nosepiece. Although these objectives should be parfocal a given microscope may be in need of adjustment. Changing from intermediate to high power should be

done by slow and careful turning of the nosepiece, watching from the side. If it is known that the objectives are not parfocal or if it appears that the high-power objective may strike the coverslip, elevate the barrel of the microscope before turning the nosepiece and then lower the highpower objective to a safe distance above the preparation. (3) Attempting to use high-power objectives on thick whole mounts. (4) Applying pressure to coverslips with stage clips or fingers. Many preparations used in the course are delicate and are easily damaged in this way. Stage clips should rest only upon the ends of the slide, not upon the coverslip.

Attention is called also the importance of clean lenses and slides. Oculars and objectives should be cleaned with lens paper according to directions given by the instructor. Dust and finger marks may be removed from a slide by very gentle wiping with a cloth after fogging the surfaces of the slide with the breath. Instructions on the correct use of mirror, substage condenser, and source of light will be given by the instructor.

Identify the following structures in the preparation of the rat ovary.

Germinal epithelium: the epithelial covering of the ovary, consisting of a single layer of low cuboidal or squamous cells. See figure 3.

Tunica albuginea (L. *tunica* = mantle; *albuginea* = white): a narrow, dense, fibrous layer of connective tissue immediately beneath the germinal epithelium.

Stroma (G. *stroma* = bed): the interstitial connective tissue of the ovary, consisting of a network of connective tissue fibers and spindle-shaped cells.

Follicles: nests of cells, each enclosing a developing germ cell.

Primary follicle: a small spherical or oval nest of cells, underneath the tunica albuginea, frequently where the

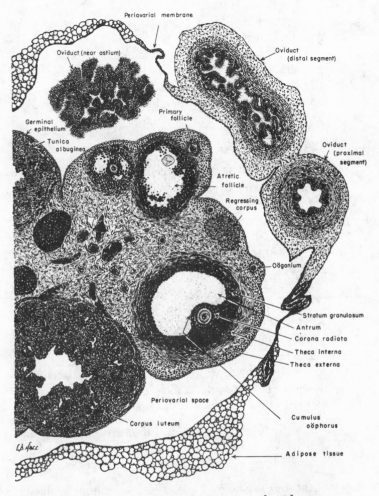

Fig. 3. Sectional view of rat ovary and oviduct.

surface of the ovary is indented; consists of the following structures.

Follicle cells: a single layer of low cuboidal or flattened cells with dark nuclei surrounding an oögonium.

9

Oögonium (G. *oion* =egg + *gonos* = producing): a potential ovum, much larger in size than the follicle cells, possessing a large vesicular nucleus with a loose network of chromatin and a prominent nucleolus. Small oögonia, unsurrounded as yet by follicle cells, may also be found in the tunica albuginea or cortical stroma.

Growing follicles: follicles of varying size, differing from primary follicles, as follows.

Follicle cells: several layers of cuboidal cells.

Antrum (G. *antron* = cavity): cleft or cavity of varying size formed within the layer of follicle cells by coalescence of intercellular spaces.

Liquor folliculi: fluid content of the antrum; secreted by the follicle cells; seen in the preparation as a light pink, granular coagulum.

Oögonium or *primary oöcyte* (depending upon stage of growth of the germ cell): Oögonia become primary oöcytes at the conclusion of their growth phase. Not easily distinguished.

Zona pellucida (L. *zona* = girdle; *pellucida* = transparent): thick, white or greenish envelope separating the germ cell from the follicle cells; probably secreted by germ cell and follicle cells together.

Mature or *Graafian follicles* (after Regnier de Graaf, Dutch physician of the seventeenth century): very large follicles, bulging from the surface of the ovary.

Antrum: expansive cavity filled with liquor folliculi.

Cumulus oöphorus (L. *cumulus* = a heap; G. *oion* = egg + *phoros* = bearing): a hillock of follicle cells, protruding into the antrum, and containing the germ cell. If a given section does not show the cumulus, it is because the plane of section has passed to one side. By searching, a favorable section can be found.

Primary or *secondary oöcyte* (depending upon stage of oögenesis): The germ cell is a primary oöcyte before the first maturation division, a secondary oöcyte after this division and before the second meiotic division. A secondary oöcyte with adjoining polar body is rarely seen. The germ cell in even the largest graafian follicle seen here is probably a primary oöcyte. Very soon after the first maturation division the germ cell is ovulated (released from ovary) as a secondary oöcyte in which the second meiotic division has progressed to the metaphase stage. Oögenesis is not completed unless fertilization occurs. Find a section passing through the large nucleus of the germ cell. Note the heavy nuclear membrane, granular chromatin, and large nucleolus; the cytoplasm contains fine granules of yolk.

Stratum granulosum (L. *stratum* = layer; *granulosum* = granulous): the thick, stratified epithelium of follicle cells bordering the antrum, exclusive of the cumulus oöphorus. Early histologists mistook the nuclei of the cytoplasmic poor follicle cells for granules, hence the name stratum granulosum.

Corona radiata (L. *corona* = crown; *radiata* = radiating): the inner layer of columnar follicle cells of the cumulus oöphorus immediately adjacent to the zona pellucida. Upon ovulation this layer escapes with the secondary oöcyte remaining attached to the zona pellucida and appears as a crown, hence the name corona radiata.

Theca (G. *theke* = case) or *capsule:* two-layered envelope enclosing the follicle.

Theca interna: inner layer of the capsule immediately outside the stratum granulosum, consisting of a layer of large cells possessing oval nuclei and fine network in the cytoplasm. Capillaries and a matrix of

11

connective tissue fibers may be seen also. The follicular hormones, collectively called estrogen, are probably secreted by the theca interna.

Theca externa: outer layer of the capsule made up of spindle-shaped cells and connective tissue fibers circularly arranged. Larger blood vessels occur in this layer; the vessels may be filled with blood cells, which appear as dark, almost solidly stained bodies.

Atretic follicles° (G. *atretos* = not perforated): unovulated, degenerating follicles identified by vacuolated, disorganized stratum granulosum, free follicle cells in the antrum, and remnants of the germ cell in the form of a single shrunken cell or pale, fragmented bodies.

Corpora lutea (L. *corpus* = body; *luteum* = yellow): large, usually solid bodies, more lightly stained than follicles. Produced in sets (average of 5 per ovary at each ovulation in the rat); sets will survive several ovarian cycles, so that an ovary may exhibit many corpora of different ages. Identify the following.

Lutein cells: large, polyhedral cells containing clear, faintly vacuolated cytoplasm and large, spherical nuclei with chromatin network and one or more nucleoli. By microchemical methods a yellow lutein pigment can be demonstrated in the lutein cells. These cells have originated mostly from the stratum granulosum of the ovulated graafian follicle.

Capillaries: network of vessels, with radiating pattern, identified as empty spaces with a thin endothelial lining or by dark masses of blood corpuscles. Not always clearly seen.

° Structures which are starred are usually difficult to observe or may be regarded as of less importance.

Capsule: a layer, varying in thickness, enclosing the corpus luteum, consisting principally of fusiform cells and connective tissue fibers (both derived from the theca externa of the follicle) and groups of luteinlike cells (derived from the theca interna of the follicle) and blood vessels.

Types of corpora

Young corpora: * identified by a large central cavity. This type will probably not be seen.

Active corpora: large, highly vascularized bodies, showing healthy appearance of lutein cells; determined accurately by special histological methods.

Regressing corpora: vary from large corpora with degenerating centers to small bodies of poor organization.

Large blood vessels: identified by dark masses of blood cells within the vessels, some of which are cut crosswise and some longitudinally. Distinguish between thick-walled arteries and thin-walled veins.

Periovarial membrane (G. *peri* = all around): folded layer of peritoneum enveloping the ovary and attached to the oviduct; consisting of squamous epithelial cells and adipose tissue, identified by the large clear fat cells.

Periovarial space: space between ovary and periovarial membrane. In the rat the periovarial space does not open into the body cavity.

Oviduct: Sections of the convoluted oviduct may appear alongside of the ovary.

Ovarian (distal) segment: segment nearest to the opening of the oviduct into the periovarial cavity; lumen appearing as a labyrinth of spaces and partitions lined with distinctly ciliated columnar cells; epithelium surrounded by a thin layer of smooth muscle and connective tissue.

13

Uterine (proximal) segment: part nearest uterus, characterized by thick walls composed of a relatively smooth, pseudostratified columnar epithelium, a layer of smooth muscle fibers, arranged circularly and longitudinally, and an external layer of connective tissue.

2. PRIMATE OVARY: Examine under intermediate and high power the slide of monkey or human ovary. Compare with the ovary of the rat. Note the following.

Graafian follicles: The cortex of the primate ovary has a very large number of follicles. Distinguish between primary, growing, and approaching mature follicles; review the structure of a follicle.

Atretic follicles: degenerating follicles, especially numerous in the primate ovary. Some exhibit a large necrotic germ cell, devoid of corona and floating in the follicular liquid, and a rapidly degenerating stratum granulosum. Observe rounded follicle cells within the cavity of the follicle. In other follicles showing advanced atresia the germ cell may be completely gone or represented by a small, deeply stained elliptical body. The zona pellucida is very resistant and remains as a collapsed shell. The stratum granulosum has disappeared almost entirely (a few follicle cells may be seen in the cavity of the follicle) and has been replaced by the greatly thickened, highly vascularized theca interna, composed of epithelioid cells similar to lutein cells. The atretic follicle thus resembles a regressing corpus luteum (see below). Ultimately the atretic follicle is invaded by connective tissue and is destroyed. Observe that nests of theca cells may remain scattered in the stroma of the ovary; they are designated *interstitial glands* of the ovary.

Corpora lutea: A large, active corpus will probably not be seen. Most of the corpora are in various stages of regression. The primate corpus differs from that of the rat in having a large cavity, which in the degenerating corpus is filled with loose connective tissue and remnants of lutein cells. The

14

walls of the regressing corpus are composed largely of theca interna cells, similar to those in the atretic follicle. The atretic follicle is distinguished from the regressing corpus by the presence in the former of the collapsed zona pellucida and traces of the germ cell; unless serial sections of the ovary are available this distinction may not be made. The regressing corpus is eventually replaced by connective tissue and is known then as a *corpus albicans* (white body).

Stroma and *tunica albuginea:* Note that these structures are particularly well formed in the primate ovary.

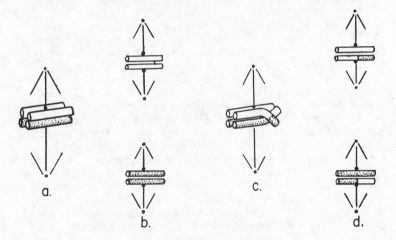

Fig. 4. Meiosis. *a.* Bivalent or tetrad in first maturation division, without crossing-over; *b,* dyads in second maturation divisions; *c,* bivalent or tetrad in first maturation division with crossing-over; *d,* dyads in second maturation divisions.

MEIOSIS

Because meiosis is so frequently misunderstood with respect to chromosomal arrangement and distribution a brief explanation is given here. In synapsis, which occurs in the prophase of the first maturation division, paired homologous chromosomes, each split lengthwise, form a bivalent (also called tetrad) consisting of four

15

chromatids, two of which (white in fig. 4, *a*) are of maternal origin and two of which (stippled) are of paternal origin. Let us assume that in the anaphase of the first meiotic division the maternal chromatids migrate to the pole of the spindle to which they are connected, the paternal chromatids to the opposite pole. In this simplified picture, in which genetic crossing-over does not occur, one might say that the first meiotic division is reductional since maternal and paternal chromatids are separated (segregated). The second division would be purely equational, the sister chromatids (also called a dyad) of a chromosome separating from each other (fig. 4, *b*). Incidentally, the tetrad (fig. 4, *a*) does not separate along the other plane so that one paternal and one maternal chromatid go to each pole. The reason for this apparently is that the two maternal chromatids (likewise the two paternal chromatids) are held together by a centromere or kinetochore, represented by the dot on the chromatids at the end of the attachment fiber. Electron mircoscopy shows that actually the centromere is already divided but the halves stick together and function as one attachment point.

We know from the genetic phenomenon of crossing-over, which probably occurs universally, that there is an exchange of segments between maternal and paternal chromatids in the first maturation division. When a bivalent is carefully examined in the metaphase stage ones sees that the opposing pairs of chromatids seem stuck together at one or more points. Because of the χ-like appearance of these points they are called chiasmata (singular = chiasma; G. *chiazein* = to mark with a X). At a chiasma the homologous chromatids have broken and then fused together so that there is a cross-connection between the two chromatids resulting in an interchange of segments (see fig. 4, *c*). In the anaphase stage the homologous chromatids pull apart (segregate). Figure 4, *d* shows the halves of a bivalent (dyads) ready for the second maturation division. Now it will be apparent, that, owing to an exchange of parts both dyads consist of two nonidentical chromatids instead of two sister (identical) chromatids. The second meiotic division will then involve a segregation of paternal and maternal material. Thus, in the light of crossing over, the terms "reductional" and "equational" divisions become somewhat meaningless. The significance of meiosis lies in the following: first, the reduction of the diploid number of chromosomes to the haploid number; second, the orderly distribution to the gametes of one representative of each type of chromosome; and third, the varied genic composition of the chromosomes made possible by crossing-over.

Schematic Outline of Changes in the Reproductive Organs of the Rat in the Estrous Cycle

Stage	Living animal	Histology of vaginal mucosa	Uterus	Ovary
I Proestrus (12 hrs.)	Vaginal mucosa slightly dry. Smear of epithelial cells only. Lips slightly swollen. In heat toward end	Cornified layer *under* surface layer of epithelium. Epithelium thick	During stage 1 uterus becomes distended with fluid increasing the diameter	Follicles rapid growing
II Estrus (12 hrs.)	Vaginal mucosa dry and lusterless. Smear of few cornified cells. Lips swollen. In heat. Copulation	Cornified layer well formed and on the surface. Epithelium thick	Reaches greatest distention and thinness of epithelium	Follicles largest. Germ cells undergo maturation
III Early metaestrus (15 hrs.)	As in stage II, but cornified material abundant in smears. Animal not in heat	Cornified layer shredding and finally completely detached	Epithelium undergoing vacuolar degeneration	Ovulation
III Late metaestrus (6 hrs.)	Vaginal mucosa slightly moist. Smear of cornified cells and many leucocytes. Swelling of lips gone	Cornified layer gone. Epithelium thin. Many leucocytes	Some vacuolar degeneration but also regeneration	Young corpora lutea. Eggs in oviduct. Follicles smallest (next crop)
IV Diestrus (57 hrs.)	Vaginal mucosa moist, glistening. Smear of leucocytes and epithelial cells	Epithelium thin. No cornified layer yet. Some leucocytes	Epithelium undergoing regeneration	Follicles of various sizes. Corpora lutea continue to grow

Modified from Long and Evans, *The Oestrous Cycle in the Rat*. Memoirs of the University of California, volume 6.

17

According to one theory, formerly in better standing than nowadays, germ cells come from the gonadal epithelium. This idea is embodied in the name, germinal epithelium, given to the covering of the ovary, as noted above. In the male a germinal epithelium does not enclose the testis, but the seminiferous tubules arise in the embryo from ingrowths of germinal epithelium, called sex cords. Adult ovarian follicles originate from proliferations of the germinal epithelium, termed ovigerous cords. Each contains a central cell, an oögonium. According to the theory the oögonium is not specialized at the outset of follicular development; it is like any other cell in the germinal epithelium. Its fate to become a gamete is merely the consequence of its position in the ovigerous cord. This idea is consonant, of course, with the basically regulatory nature of the vertebrate embryo (see p. 42).

Another theory receiving increasing support holds that presumptive germ cells are set apart from the somatic cells very early in the development of a vertebrate, as in many invertebrates. These precociously segregated elements, designated primordial germ cells, constitute a germinal line from which all prospective gametes are derived. The primordial germ cells appear to differentiate in the entoderm and to migrate by means of pseudopodia into the cortex of the gonadal ridge. They may be recognized or identified by their morphology (large size, polymorphic shape, lightly staining cytoplasm, and giant nuclei), by certain physiological characteristics (e.g., high content of alkaline phosphatase), and by some genetic markers (e.g., W and W^v genes in mice). In the bird the primordial germ cells arise in a crescentic area of extraembryonic entoderm anterior to the embryo. If this region of the blastoderm is surgically removed before the germinal cells migrate into the embryo sterility results.

Migration of the avian primordial germ cells (PGC) from the extraembryonic crescent is by the bloodstream. These cells can be identified in smears of blood taken from the dorsal aortae of chick embryos. They are very large (20-25 microns in diameter) in comparison with surrounding blood cells, and they are heavily laden with glycogen, as shown by their staining with the periodic acid-Schiff reaction (PAS). Prior treatment of a smear with diastase gives a negative PAS reaction. The cells also contain vacuoles filled with lipid. The migration of PGC begins at about the 16-somite stage, reaches a peak at 24- to 27-somite stage, and by the stage of 30-36 somites they are rarely found in blood smears. Recent studies have shown that PGC will spontaneously leave fragments of chick germinal crescent in the absence of blood vessels and migrate at random in the culture chamber. It is claimed that the PGC are chemically attracted to the presumptive gonads. If the germinal crescent and embryonic gondal ridge are cultured together the PGC migrate from the former into the latter, and if the germinal crescent of a primitive streak-stage is transplanted into the coelom of a 30-somite embryo PGC may be observed migrating from the graft through the epithelium of the gonadal ridge. The number of PGC in vertebrate embryos averages between 20 and 100 per embryo—a relatively small number of cells, and they are remarkably uniform in their large size.

The uncertainty in drawing the conclusion at the present time that gametes in vertebrates are only derived from primordial germ cells is due to the fact that these cells have not been identified in the germinal epithelium of the adult ovary. Until future research discloses them one can only assume that they are present, that they have descended from the large entodermal cells seen earlier, and that although resembling neighboring somatic cells morphologically they possess distinctive biochemical and physiological features and a specific developmental potentiality.

One can push the ontogenetic history of the PGC in some forms to still earlier developmental stages, indeed to the zygote and even to the oöcyte. There is a "germinal cytoplasm" in amphibian eggs in the form of small islands of cytoplasm immediately beneath the cell membrane in the vicinity of the vegetative pole. This plasm is distributed by cleavage to the vegetal blastomeres which later become entodermal cells in the course of gastrulation. The germinal cytoplasm is rich in small yolk platelets, pigment, and mitochondria. The islands contain RNA. Irradiation of the vegetal hemisphere of zygotes with ultra-violet light causes partial or complete sterility. It is tentatively concluded that the UV-sensitive material (RNA ?) is responsible for the germ line.

Review Questions

Questions placed at the end of each section of the syllabus are designed to enhance the significance of the laboratory exercise just completed.

1. Define each of the following terms: mitosis, meiosis, metaphase, homologous chromosomes, chromatid, synapsis,

crossing over, primary oöcyte, second polar body, oötid, haploid, sex chromosomes, autosomes, nucleolus, bivalent, tetrad, dyad.

2. In man, how many chromosomes are there in each of the following cells (for purposes of this exercise regard each chromatid of a tetrad or dyad as a separate chromosome): zygote, oögonium, metaphase of a dividing liver cell, metaphase of a primary oöcyte, first polar body?

3. Diagram oögenesis. Begin with an oögonium possessing six chromosomes; distinguish between maternal and paternal chromosomes by colored pencils; show distribution of the chromosomes in the two meiotic divisions.

4. What probable physiological exchanges occur between a developing germ cell and the surrounding follicle cells?

MEMBRANES

The last question above prompts a review of the nature of cell and nuclear membranes. Membranes are from 60 to 100 A in width and consist of a bilayer of lipid (i.e., two layers, each one molecule thick). Protein particles lie in the inner, outer, or both lipid layers. Some protein particles extend from internal to external surfaces of the membrane (transmembranous particles). The membrane should be thought of as fluid and dynamic, as protein particles appear to be easily moved. Some membranes are structural (providing support), others are enzymes (catalyzing chemical reactions in and on the membrane); some facilitiate the movement of substances through a membrane (e.g., facilitated diffusion and active transport), others perform special functions (e.g., absorption of light). Perhaps all membranes have an external coating of glycoproteins (protein conjugated with carbohydrate), and glycolipids (protein conjugated with lipid) that provide important binding sites which "recognize" and hold molecules. Certain membranes have pores through which relatively large molecules pass. The pores in the nuclear envelope— a double membrane separated by a narrow space—are especially large (from 300 to 1000 A in diameter). Other structures should no longer be termed membranes (e.g., vitelline membrane, fertilization membrane, basement membrane), as they are not true membranes but layers, or coats, or lamina.

B. Vaginal and Uterine Cycles in the Rat

1. VAGINAL CYCLE: The cyclical changes in the vagina, although relatively slight in primates, are quite marked in the rat and other rodents and are correlated with the ovarian cycle just described and the uterine cycle to be considered later. In the living animal the condition of the vaginal epithelium is diagnosed by an examination of the vaginal contents obtained by washing or gently scraping the vaginal mucosa with a small spatula. The histological makeup of the vaginal fluid varies in the course of the cycle (see table) and reflects the changing histological picture of the vaginal epithelium. This simple technique (vaginal smear) permits a rapid and accurate determination of the stages in the estrous cycle.

Examine under intermediate and high powers the longitudinal sections of the rat vagina. Each section illustrates a different stage in the vaginal cycle.

Stage I. Proestrus (L. *pro* = before + G. *oistros*). See figure 5, *a.*

> *Surface layer:* a well-defined stratum, three to five cells in thickness, composed of lightly stained and loosely arranged cells. In places, it may be noted that this layer is becoming separated from the deeper layers of the mucosa and is being cast off into the cavity of the vagina. Vaginal smear at proestrus shows many epithelial cells.

> *Stratum corneum* (L. *corneum* = horny): pink (acidophilic) layer immediately under the surface layer, consisting of densely packed, squamous, cornified (keratin-containing) cells which are dead and no longer show nuclei or cell boundaries.

> *Stratum granulosum:* narrow layer, one to three cells in thickness, below the stratum corneum; consists of large, highly granular (basophilic) cells.

21

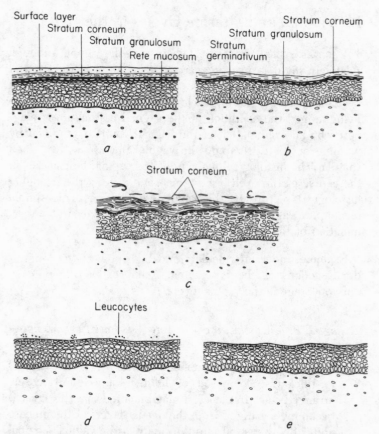

Fig. 5. Selected histological pictures of the vaginal cycle in the rat. *a*, Proestrus; *b*, estrus; *c*, early metaestrus; *d*, late metaestrus; *e*, diestrus.

Rete mucosum (L. *rete* = net; *mucosum* = slimy): several layers of nongranular polyhedral or cuboidal cells situated beneath the stratum granulosum.

Stratum germinativum (L. *germinare* = sprout): layer of columnar cells which, by mitotic divisions, generate the more superficial layers.

22

Stage II. Estrus. See figure 5, *b.*

Surface layer: no longer present.

Stratum corneum: at the surface now; thick, well developed; beginning to split and shred. Vaginal smear at estrus shows some cornified cells.

Stage III. Metaestrus (G. *meta* = after + *oistros*). See figure 5, *c* and *d.*

Early phase: stratum corneum sloughing in large shreds; already cast off in lower part of vagina. Vaginal smear in early metaestrus shows masses of cornified cells.

Late phase: characterized by thin epithelium, highly infiltrated with leucocytes. Stratum corneum and stratum granulosum are absent; nucleated squamous cells are on the surface. Note leucocytes in the cavity of the vagina. Vaginal smear in late metaestrus shows many leucocytes with perhaps a few cornified and noncornified epithelial cells.

Stage IV. Diestrus (G. *dia* = through or between + *oistros*). See figure 5, *e.* Characterized by thin epithelium. Vaginal smear in diestrus shows some leucocytes and a small number of squamous epithelial cells.

2. UTERINE CYCLE: The cyclical changes in the histology of the rodent uterus are not marked, as they are in primates. Examine the cross sections of the rat uterus.

Stage I. Proestrus: in this stage uterus distended with fluid. The uterine epithelium consists of a single layer of simple columnar cells.

Stage II. Estrus: now uterine distention maximal. Uterine cavity may reach 5 millimeters in diameter. Uterine epithelium reduced in height to simple low columnar or cuboidal. Blood vessels filled with blood (vascular congestion).

23

Stage III. Metaestrus:

Early phase: uterine cavity collapsed; epithelium high columnar, pseudostratified. Vacuolar degeneration may be observed in the distal parts of the cells. The time of greatest uterine impairment thus agrees with the time of greatest epithelial sloughing in the vagina. There is, however, no dehiscence of the uterine epithelium in the rat.

Late phase: uterine cavity collapsed; leucocytic infiltration in epithelium marked; leucocytes present also in uterine cavity; vacuolar degeneration and regeneration both take place at this time.

Stage IV. Diestrus: uterine lumen slitlike; no leucocytic infiltration; epithelium simple columnar.

MENSTRUAL CYCLE

In primates the estrous cycle is characterized by monthly uterine bleeding, hence the primate cycle is more specifically designated menstrual (L. *mensis,* month) cycle. Figure 6 is a highly schematized representation of the human menstrual cycle with or without a state of pregnancy following. Events in the uterus are correlated with those in the ovary and both with time in days, indicated on the horizontal axis. Shown also are curves, some hypothetical, of the blood concentration of pituitary and ovarian hormones known to be involved in the estrous cycle. Indeed the causation of these structural and physiological changes in the female mammal is the interplay of these hormones and others, possibly including some not yet discovered. Because the basic control of the cycle is hormonal and because there is variation in the relative concentrations of these substances, the cycle is not a precise schedule of events. Therefore, the dates given in the figure represent only the statistical average time of occurrence, such as: antrum formation at 7 days, ovulation at 14 days, implantation at 20 or 21 days, mensis at 28 days, and parturition at 280 days.

At least three pituitary gonadotrophic hormones, produced by the anterior lobe, are involved. (1) Follicle-stimulating hormone (FSH), also called gametokinetic hormone, stimulates growth and maturation of the follicle, especially the germ cell and stratum granulosum. (2) Luteinizing hormone (LH), also termed inter-

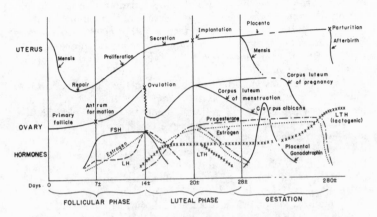

Fig. 6. Diagram of human pregnant and nonpregnant cycles showing sequence of changes in ovary and uterus and relative blood concentrations of pituitary, ovarian, and placental hormones.

stitial cell-stimulating hormone (ICSH), stimulates the theca interna, induces corpus luteum formation, and acts synergistically with FSH in bringing the graafian follicle to maturation and in causing ovulation. (3) Luteotrophin (LTH), also designated lactogenic hormone or prolactin, maintains the corpus luteum and stimulates the secretion of progesterone by the corpus. This hormone also stimulates lactation by the mammary gland after the birth of the fetus. Other pituitary hormones may be involved in the regulation of the estrous cycle and in parturition. The posterior lobe hormone oxytocin, is probably important in initiating the process of labor.

At least two ovarian hormones play a role in the estrous cycle. (1) Estrogen, a collective term for follicular hormones, promotes the proliferation phase of the uterine cycle, stimulates other female structures such as oviducts, vagina, and mammary glands, and over a long period of time affects the secondary sex characters such as the skeletal, muscular, adipose, and epidermal features of the female body. Even certain aspects of behavior in some mammals are influenced by estrogen. The follicle secretes estrogen in the first half of the cycle, the corpus luteum in the second half. (2) Progesterone, produced by the corpus luteum, promotes the secretion phase of the uterine cycle and stimulates other structures such as the mammary gland. Both estrogens and progesterone affect the

25

pituitary, stimulating the release of some gonadotrophins, inhibiting the production of others—hence the picture of a complex interplay of gonadal and gonadotrophic hormones. Other chemical mediators are more generally involved in these physiological processes and structural changes in the female mammal such as thyroxin, adrenal hormones, relaxin, and so on.

The placenta has important endocrine functions in addition to its role in fetal nutrition, respiration, and excretion. Estrogen and progesterone are released in large amounts by the placenta, as shown in the figure, until near the close of gestation when for some unknown reason placental failure occurs and leads to a drop in their concentration and to the onset of birth processes. In addition, the placenta produces an anterior pituitarylike hormone called placental (or chorionic) gonadotrophin. This hormone is excreted by the pregnant female and a positive pregnancy test is based on the demonstration of this substance in the urine.

A recent review (Jaffee, R. B. and Midgley Jr., A. R., *Obstetrical and Gynecological Survey*, 24:200-213, 1969) of human gonadotrophin radioimmunassay presents curves for follicle-stimulating hormone and luteinizing hormone obtained by averaging the data on menstrual cycles in 37 women, ages 18-32. The technique employed assays serum concentrations of FSH and LH. It is based upon the principle of competition between radioactively labelled hormone and unlabelled hormone for a limited amount of antibody. The data give essentially the following picture (fig. 7, modified from Jaffee and Midgley).

In agreement with the highly schematized diagram (fig. 6) one notes that FSH and LH reach serum concentrations at the midpoint of the cycle (resulting in ovulation) and that the concentrations of both gonadotrophins fall thereafter. The new curves differ from the diagram in the following features: through the cycle except for the period of mensis, the peaks for both hormones are sharp instead of arched, and the lowest concentrations of both hormones occur just before menstruation. Observe that the average curve of basal temperatures in the 37 women shows a rise in temperature only after the LH peak. The rise is gradual and requires five or six days to reach or exceed 98° F. Are basal temperature curves useful in population control? The 37 normal cycles studied averaged 29.9 days with a range of 24 to 38 days. The LH peak was at 17.7 days (mean). The serum concentrations of FSH and LH in males are about the same as those at the end of the cycle in women. FSH in the male is slightly higher than LH. There is no evidence of rhythmicity or regular peaking in samples obtained once a day for three to ten weeks in four men.

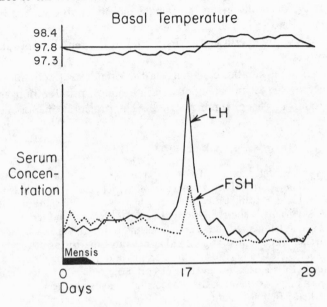

Fig. 7. Curves for FSH and LH and basal vaginal temperature in the human menstrual cycle based upon radioimmunassay of gonadotropins in the blood (From Jaffee and Midgley).

Review Questions

1. What is a tissue? Name the major tissues of the body. Give the subtypes of epithelium, muscle, and connective and supportive tissue.

2. List the layers of thick epidermis such as that from the heel of man. Compare these layers to the strata of the vaginal mucosa in the rat.

3. Give the site of production, function, and period of action in relation to the menstrual cycle of the following hormones: a) follicle-stimulating hormone (FSH), b) luteinizing hor-

mone (LH), c) estrogen, d) progesterone, e) luteotrophin, f) placental gonadotrophin.

4. What conditions in the female reproductive tract are unfavorable to the survival of the sperm?

5. How are sperm cells transported in the female reproductive tract?

6. Define: keratin, collagen, elastin, pseudostratified, anovulatory cycle, estrus, gonadotropic hormones, proliferation and secretion phases of the menstrual cycle, prolactin, menopause.

PARTURITION

It is increasingly apparent that the fetus plays a role in initiating separation (parturition) from its mother. The fetal hypothalamus (ventral lateral wall of diencephalon) first responds to some stimulus (stress perhaps) shortly before the end of normal gestation (see p. 239). It triggers the anterior lobe of the pituitary to release ACTH (adrenocorticotropic hormone) that activates the cortex of the fetal adrenal gland to secrete corticosteroids and estrogen. One of the placental responses to elevation of corticosteroids is production and release of prostaglandins (20-carbon fatty acids synthesized by the deciduae) that stimulate the myometrium (smooth muscle layer) of the uterus to begin contractions (labor). Increased levels of estrogen from adrenal gland and chorion (syntrophoblast) facilitate the contractions by reducing the membrane potential (to -50 mv) of the muscle fibers. At the same time there is a decrease in progesterone. This hormone normally inhibits myometrial activity by maintaining a high membrane potential (-65 mv). Accordingly, the fall in progesterone concentration reduces the inhibition to labor. At the same time, the fetal hypothalamus synthesizes the neurohormone oxytocin (an octapeptide) which is released from the posterior lobe of the pituitary. Oxytocin also stimulates the myometrium, whose sensitivity to the hormone is enhanced by estrogen and depressed by progesterone. Finally, there are maternal factors inducing myometrial contractions: maternal oxytocin and stimulation by the maternal sympathetic nervous system. Both the hormonal and neural pathways begin, as in the fetus, in the hypothalamus. What activates the hypothalamus? Probably afferent impulses initiated by mechanical stimulation of uterus, cervix and vagina owing to the increase in volume of uterine content.

C. Spermatogenesis in the Mammal

1. RAT TESTIS: Whereas in the female the sex cells mature periodically in estrous cycles, in the male the spermatozoa are produced throughout certain seasons of the year (rut) or, in primates and domestic mammals, continuously. Not all parts of an active testis will, however, show mature sperm cells at any given time. Spermatogenesis occurs in overlapping waves which pass along the tubules of the testis so that the stages of spermatogenesis observed at any given level of a tubule will be different from those at another level. In the rat the length of a spermatogenic wave has been found to be about 32 mm. along the tubule. The wave moves slowly. The estimated time for spermatogenesis in the rat is twenty days. Examine the rat testis at intermediate and high magnification. Identify the following structures.

Tunica albuginea: thick capsule about testis, composed of collagenous fibers and flattened, elongated nuclei of connective tissue cells (fibroblasts).

Seminiferous tubules: large number of thick-walled tubules cut at various angles, crosswise, longitudinally, and obliquely.

Interstitial cells: clusters and strands of cells situated between the seminiferous tubules. Examine the cells under high power; note relationship to blood vessels, especially capillaries. These cells secrete the male sex hormone.

Blood vessels: Identify capillaries, small arterioles or venules, and large arteries or veins.

Spermatogenesis: Any given cross section of a seminiferous tubule will show three or four overlapping spermatogenic waves. The oldest wave is represented by cells bordering the lumen of the tubule. These cells are either maturing spermatozoa or metamorphosing spermatids. The youngest wave, just beginning, is represented by the germ cells near the outer surface of the tubule. These cells are the spermatogonia ready to undergo meiosis. Waves of intermediate age are represented by the cells in the middle layer of the tubular wall. Any given cross section may usually be found to approximate one of the following histological pictures, referred to as examples A to E.

Example A (see figs. 8 and 9, *a*): the histological picture most commonly found. Observe under high magnification the following features.

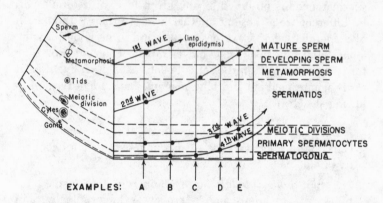

Fig. 8. Schematic representation of a sector of a seminiferous tubule in the rat showing four spermatogenic waves as seen in five different cross-sections (examples A to E).

Basement membrane: a very thin fibrous layer encapsulating each tubule. Note the flattened, elongated nuclei of fibroblasts widely spaced apart and lying within or on the outside of the membrane.

Spermatogonia (G. *sperma* = seed + *gonos* = producing): single layer of cells beneath the basement membrane, possessing darkly stained, oval nuclei.

Sertoli (sustentacular) cells (after E. Sertoli, Italian histologist): larger cells interspersed among the spermatogonia, identified by large, round, or three-sided nuclei lightly stained and possessing a large nucleolus.

Medium primary spermatocytes: a layer of cells inside the spermatogonia described above. The nuclei are larger and more deeply stained than those of the spermatogonia (true for example A only). The dark chromatin structures are synapsed chromosomes. Primary spermatocytes represent former spermatogonia which have passed through a growth phase, migrated one step toward the cavity of the seminiferous tubule, and begun nuclear changes in preparation for the first meiotic division.

Spermatids (L. & G. *id* = suffix meaning daughter of): four or five rows of cells lying inside the primary spermatocytes. The nuclei are round, relatively clear, with a lightly stained chromatin network. These cells, belonging to an earlier spermatogenic wave, have undergone both meiotic divisions and are soon to metamorphose into mature sperm cells.

Developing spermatozoa: identified by the elongated, solid black sperm heads which are arranged in groups; each group embedded in the cytosome of a Sertoli cell, which stretches as a narrow process from the nucleus of the cell to almost the cavity of the tubule. From the hook-like head of the developing sperm there extends toward the lumen of the tubule the thick, elongated cytoplasmic part of the cell, from the inner end of which emerges into the lumen the sperm tail or flagellum. These maturing spermatozoa belong to a still earlier spermatogenic wave.

Example B (see figs. 8 and 9, *b*):

Mature spermatozoa: The spermatozoa approaching maturity in example A are now completely formed and lie in a row at the edges of the lumen with their tails centrally directed. The sperm heads are no longer attached to the Sertoli cells, which have contracted, and are separated

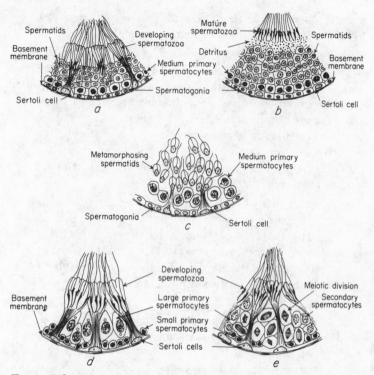

Fig. 9. Selected histological pictures of spermatogenesis in the rat. (From Lenhossék, Arch. f. mikr. Anat., 51:215.) *a–e*, Examples A to E (see fig. 8).

from the spermatids by a layer of detritus consisting of chromophilic granules and globules. Observe in detail the form of a mature sperm cell.

Layer of detritus: the layer of dark granules, mentioned above, formed by cytoplasm sloughed from metamorphosing sperm cells and from Sertoli cells.

Spermatids: no change in this spermatogenic wave; same as in example A.

Medium primary spermatocytes: Note that nuclear size has increased; synapsed chromosomes are shorter and more visible; an occasional primary spermatocyte has moved in among the spermatids and toward the lumen of the tubule.

Spermatogonia: as in example A.

Example C (see figs. 8 and 9, *c*):

Spermatozoa: generally absent, although a few may be seen in the lumen of the tubule. Mature sperm observed in example B have left the tubule, thus completing the oldest spermatogenic wave being followed. The sperm pass through a short set of tubules, the *rete testis* and *efferent ductules,* and into the *epididymis* where they mature physiologically. They are moved presumably by pressure of secretion products of which they themselves are a part.

Metamorphosing spermatids: Spermatids seen in examples A and B, representing the next oldest spermatogenic wave being followed, are now in an early stage of metamorphosis (spermiogenesis). Their nuclei are no longer round and centrally placed in the cell but are oval and situated at the border of the cell nearest the outer surface of the seminiferous tubule. Nuclei still show faint chromatin granules. The cytoplasm of each spermatid, like an appendage of the nucleus, is an oval body oriented toward the lumen into which have grown the flagella of the future sperm cells.

Medium primary spermatocytes: The layer of developing spermatocytes observed in examples A and B and constituting the third spermatogenic wave being followed now exhibits further changes leading to the first meiotic division, namely, additional increase in nuclear size and greater condensation of the chromosomes which have become duplicated thus forming tetrads (not clearly shown). Some have taken "another step" toward the lumen of the tubule.

Spermatogonia: beginning to enlarge, to stain more deeply, and to move toward the adjoining layer of primary spermatocytes. A *new* spermatogenic wave is just beginning. Not all spermatogonia pass through stages leading to spermatocytes. Some remain as resting or stem spermatogonia for future waves. They require special preparations for identification, however.

Sertoli cells: beginning to send forth cytoplasmic processes toward the metamorphosing spermatids (difficult to observe).

Example D (see figs. 8 and 9, *d*):

Developing spermatozoa: The spermatids just beginning metamorphosis in example C show now the condensation and elongation of the nuclei to form the heads of the future spermatozoa. Their cytoplasmic bodies have become more elongated and flasklike. The developing spermatozoa are now embedded in the Sertoli cells.

Large primary spermatocytes: similar to those in example C except that nuclear size and condensation of the chromosomes now reach a maximum. The nuclei are oval with their long axes parallel to the radii of the tubule. The cells are ready for the first maturation division. Some of the spermatocytes have moved farther toward the lumen.

Small primary spermatocytes: The new spermatogenic wave begun earlier is now reflected by the transformation of the spermatogonia into small primary spermatocytes.

Spermatogonia: Resting spermatogonia are very few in number and are difficult to demonstrate in standard preparations.

Example E (see figs. 8 and 9, *e*): the most infrequently encountered picture; considerable search may be required to find a tubule showing these features.

Meiotic division: Spermatocytes in meiotic division constitute the chief feature of this picture. The division figures lie in the zone of the tubular wall between the outer layer of young spermatocytes (transition spermatogonia) and the inner layer of developing sperm cells, and are found usually in restricted regions of the tubule. That is, only certain sectors of the cross-sectional pictures of the tubule will show dividing spermatocytes. Distinctions between first and second meiotic division figures can not be made on the preparations and with the microscopic equipment at hand. A given group of dividing cells may be undergoing first *or* second meiosis; there is rarely a mixture of the two divisions.

Secondary spermatocytes:° The cells resulting from the first division; may frequently be found in sectors immediately adjoining the area of division figures. These cells may be identified by the size of their nuclei which is intermediate between that of the large primary spermatocyte (example D) and that of the spermatid (examples A and B). Moreover, the chromatin of the secondary spermatocyte is not so dense and dark as that in the primary spermatocyte but more prominent than that in the spermatid.

Developing spermatozoa: Spermiogenesis has advanced slightly over the picture in example D; the sperm heads are blacker and better formed.

Small primary spermatocytes: as in example D.

Spermatogonia: Resting spermatogonia are very few in number. By mitotic divisions their number will be increased and the picture is returned to that seen in example A.

2. RAT EPIDIDYMIS (G. *epi* = upon + *didymos* = testicle): Adjoining the rat testis just studied lies the accessory organ, the epididymis into which sperm are passively moved. Study

its structure under high and low power. Identify the following.

Ductus epididymis: a greatly coiled duct seen cut many times at all conceivable angles; lined with a high simple columnar epithelium. Outside the epithelium is a narrow layer of smooth muscle cells, circularly arranged, and a thin layer of connective tissue.

Spermatozoa: masses of mature sperm cells seen in the lumen of the ductus epididymis.

3. DEMONSTRATIONS:

Sertoli cells: Under oil immersion objective the attachment of sperm heads to the Sertoli cells may be better observed.

Human testis: Compare grossly with the rat testis just studied.

Human spermatozoa: Note head, middle piece, and tail.

Review Questions

1. Diagram spermatogenesis in mammals. Begin with a spermatogonium possessing six chromosomes; distinguish between maternal and paternal chromosomes with colored pencils; show distribution of the chromosomes in the two meiotic divisions.

2. Describe the process of spermiogenesis, including the formation of acrosome, axial filament, and mitochondrial sheath.

3. What is the function of the epididymis? How are spermatozoa moved along the duct of the epididymis?

4. What is the function of the interstitial cells? What histological evidence suggests this function?

5. What are the effects on the testis of: X-rays, vitamin E deficiency, high temperature, FSH and LH hormones?

6. What is the probable length of the functional life of the human sperm? In other animals?

7. Define: caput, corpus, and cauda epididymis; cryptorchism; heterogametic sex; sex-linked character, Golgi apparatus, bulbourethral glands, parthenogenesis.

Spermiogenesis

To the picture of spermatogenesis provided by the above study, some finer details on the metamorphosis of a spermatid (spermiogenesis) should be added. Electron microscopy has given a wealth of information on the development of sperm head and tail (flagellum).

The acrosome, a vesicle filled with granular material, forms in the Golgi apparatus and takes a terminal position in the head, beneath the cell membrane and above the nucleus. The nucleus becomes elongated and condensed, and in places the two membranes of the nuclear envelope and the cell membrane appear to fuse. Meanwhile, the centrioles migrate to the base of the spermatid. One (distal centriole) gives rise to the tail's axoneme which is a bundle of microtubules arranged in the typical ciliary pattern of a ring of nine doublets and two central singlets ($9 \times 2 + 2$). The sliding of the doublets, powered by ATP, is the basis of flagellar motility. A proximal centriole lies near, but at an angle to the distal one. The narrow beginning of the tail, containing the centrioles and other structures, is called the neck. As the axoneme extends into the lengthening tail, it becomes surrounded by nine large, dense fibers that probably also play a role in sperm motility. Mitochrondria, the suppliers of ATP, migrate into the next segment of the tail—termed the middle piece—where they line up end-to-end to form a tight helix around the fibers and axoneme. The remainder of the tail consists of axoneme and fibers, a little cytoplasm, and cell membrane. Excess cytoplasm, granules, lipid droplets and superfluous organelles are extruded from the late spermatid as a membrane-bounded body (see detritus above). Until late spermiogenesis clusters of differentiating germ cells are bound to one another by prominent cytoplasmic bridges that result from incomplete subdivisions of the cells. These bridges probably ensure synchrony in the development of batches of sperm cells.

II

EARLY DEVELOPMENT

A. Early Development in the Starfish

The echinoderm egg and embryo have been classic materials of study in embryology, earlier for descriptive work, nowadays for experimental studies. The echinoderm egg has become literally the meeting ground of embryologist, cytologist, cellular physiologist, and biochemist.

Preparations of eggs and young embryos of the starfish will be useful to review with fixed material certain features in the development of a form with a relatively small amount of yolk. The one slide is a carmine-stained whole mount of various stages of development of *Patiria miniata* from unfertilized eggs through gastrulae. Note the following.

Unfertilized egg: unactivated ovum, identified by the absence of a loose saclike enclosing membrane. The carmine preparation is unfavorable for the observation of cytological features such as nuclear apparatus and cytoplasmic organization. The apparent homogeneous granular interior, which is about all one can observe, belies the fine organization present in the living egg. The *vitelline membrane* (L. *vitellus* = yolk) and *cortical cytoplasm* are roughly represented by the narrow band at the surface of the egg. The membrane, exceedingly fine and indistinguishable from the cortex of the egg, is formed by the egg and is hence a primary membrane. Examples of secondary and tertiary membranes applied externally to the vitelline membrane will be encountered in later studies.

Fertilized egg or *Zygote:* an uncleaved cell enclosed by a *fertilization membrane,* a loose filmy investment. The fertilization membrane is formed by a breakdown of cortical plasm and the creation of a space, the *perivitelline space,* between the membrane and the surface of the egg. The vitelline membrane is now a part of the fertilization membrane. The appearance of these features is a commonly used morphological index of fertilization.

Eggs

The term "egg" is loosely used and does not always mean ovum. In many annelids, nemerteans, and molluscs the eggs are shed as primary oöcytes. The nuclei of such cells are large and are designated by the old-fashioned term, *germinal vesicle.* In some forms, the annelid *Nereis* for example, meiosis is not initiated until the sperm enters the primary oöcyte. In other types, such as the mollusc *Mytilus,* the germinal vesicle breaks down after the primary oöcyte is released but maturation proceeds only as far as the metaphase stage of first division. Meiosis will not continue until the egg is activated by sperm entrance or by some other stimulating factor. In vertebrates, as we have learned earlier, meiosis progresses as far as the metaphase of the second division—but no farther—before sperm entrance. In many echinoderms and coelenterates the eggs at the time of shedding are truly ova and the sperm enters after maturation is complete. In instances in which the sperm enters before meiosis is begun or completed the sperm head or male nucleus "waits" in the cytoplasm of the egg until the female nucleus is formed at the conclusion of the second meiotic division.

Eggs are commonly classified according to the relative amount of yolk which they contain. Those with little yolk are designated *miolecithal* (G. *meio* = small + *lekithos* = yolk): eggs of echinoderms, coelenterates, amphioxus, and mammals except monotremes. Those with moderate to considerable yolk content are termed *medialecithal* (L. *medius* = middle + *lekithos*) or *mesolecithal*: eggs of annelids, molluscs, lampreys, lung-fishes, and amphibians. Eggs with large quantities of yolk are *megalecithal* (G. *megas* = large + *lekithos*): eggs of arthropods, hag-fishes, bony fishes, reptiles, birds, and monotremes. The terms *isolecithal* (G. *isos* = equal + *lekithos*) and *telolecithal* (G. *telos* = more at one place + *lekithos*) have different connotations. The former refers to an even distribution of yolk in the egg, a feature probably not attained in any form. The miolecithal egg of the placental mammal would approach a true isolecithal condition. The yolk in most if not all ova is more concentrated at or near the vegetal pole, hence they are telolecithal. The degree varies, of course. Echinoderms are slightly telolecithal, amphibians moderately telolecithal, birds highly telolecithal. The amount of this inert stored food present within an egg has a profound influence upon the mechanics and patterns of development as we shall later note.

Cleavages: Observe 2-cell, 4-cell, and 8-cell stages of development resulting respectively from the first, second and third cleavages. Observe the size and arrangements of the cells, known as *blastomeres* (G. *blastos* = sprout + *meros* = part).

Blastula: a ball of cells, or one-layered embryo, resulting from successive cleavages. Slightly larger cells at one side mark the lower or *vegetative pole.* Is there evidence of a blastocoel (G. *blastos* + *koilos* = cavity)? When the blastula contains about a thousand cells cilia are formed, the fertilization membrane is broken, and the embryo becomes a free-swimming individual.

Gastrula (G. *gastros* = stomach): the two-layered embryo resulting from the invagination of cells at the vegetative pole of the blastula. Gastrulae may be best studied in lateral aspect. Identify the following:

Ectoderm (G. *ektos* = outside + *derma* = skin): the outer layer of cells of the gastrula; one of the three primary germ layers.

Archenteron (G. *archi* = first + *enteron* = intestine): the inner tube formed by the invaginated vegetative cells. The walls of the archenteron may be said to be mesentoderm, that is, composed of cells which will give rise to the other two primary germ layers, mesoderm (G. *mesos* = in the middle + *derma*) and entoderm (G. *entos* = within + *derma*).

Early, middle, and late gastrulae: identified by the extent of invagination as indicated by the length of the archenteron. In early gastrulae invagination is just beginning and the archenteron is short. In late gastrulae, which have become elongated, the archenteron extends over half the length of the embryo and at its apex may be seen a thin-walled vesicle. From this vesicle two lateral pouches will grow out later. These are the *coelomic sacs* which become separated from the archenteron and give rise to many mesodermal structures in the animal. The coelom of the echinoderm arising by outpocketings from the archenteron is called an *enterocoel* (G. *enteron* + *koilos*).

Blastopore (G. *blastos* + *poros* = passage): the opening of the archenteron to the outside, best seen in gastrulae oriented so as to permit an end view of the embryo. The blastopore marks the posterior end of the embryo and becomes the future anus.

Mesenchyme (G. *mesos* + *enchyma* = infusion): mesodermal cells scattered between the ectoderm and the archenteron. They arise by budding from the outer walls of the archenteron, especially from the vesicle at its anterior end.

Apical sensory plate: a thickening of the ectoderm at the anterior end of the gastrula (point opposite the blastopore and the position of the old animal pole). Observed in only the very late gastrula.

Bipinnaria larva: (L. *bi* = two + *pinna* = feather): the larva of the starfish which develops from the gastrula. The gut is completed by the formation of a mouth through an ectodermal invagination, the stomodeum (G. *stoma* = mouth + *hodaios* = on the way), which meets and opens into the archenteron. The archenteron differentiates into three regions of the gut: oesophagus, stomach, and intestine. Cilia, which covered the entire ectodermal surface of the late blastula and gastrula, become very long and concentrated into bands, which are the principal locomotor organs of the

41

echinoderm larva. Lobelike outgrowths, known as larval arms, develop symmetrically on both sides of the body giving the larva a bipinnate appearance. Examine the demonstration of the bipinnaria larva.

ECHINODERM THEORY OF CHORDATE ORIGIN

With models or text-book illustrations compare echinoderm larvae and the tornaria larva of *Balanoglossus,* regarded by some taxonomists as a chordate or an organism with chordate affinities. Herein lies one of the arguments for the echinoderm theory of chordate origin. The proponents of the theory attribute considerable importance to similarities between echinoderm and tornaria larvae, particularly in regard to symmetry, locomotor organs, digestive tract, and coelomic pouches and their enterocoelic origin.

In other respects echinoderm and chordate developments are similar. Both exhibit the same kind of cleavage (see p. 49). In both the blastopore forms the anus whereas the mouth arises from the stomodeum (deuterstomes, from G. *deuter* = second + *stoma* = mouth). In annelids, arthropods, molluscs, etc., on the other hand, the blastopore gives rise to the mouth (protostomes, from G. *proto* = before or first + *stoma*). Both echinoderms and chordates show regulatory development as indicated, for example, by the totipotence of isolated blastomeres and by a high degree of regenerative ability. *Totipotence* (L. *totus* = whole + *potens* = powerful) means that an isolated fragment of a developing system such as an early blastomere can form a whole embryo, albeit proportionately smaller in size than the normal embryo. Although this feature is characteristic of the deuterostomes, there are some exceptions (e.g., the ascidians or tunicates). The protostomes, on the other hand, tend to exhibit restricted regulatory powers and a more mosaic type of development in which the fates of districts of the developing system are irreversibly fixed early in development, in some instances in the zygote. Moreover, they show little regeneration of lost parts. Correlated with the above difference in regulation is the greater prominence of embryonic induction in the echinoderm-chordate line than in the annelid-mollusc line. By embryonic induction is meant the influence of one part of an embryo on another in determining the fate of the latter, i.e., what kind of cells, tissues, and organs the induced part will form (see p. 61).

Finally, there are other developmental similarities between echinoderms and chordates. Both derive their photoreceptors from cilia-like processes of the sensory epithelium which appear to be induced by centrioles that migrate to the surface of the cells. In protostomes, on the other hand, the photoreceptors are predominately non-ciliary or rhabdomeric in type, that is, consisting of arrays of microvilli which appear to arise independently of cilia. In both phyla the skeleton is internal and develops from mesoderm—calcareous plates in echinoderms, bones in chordates.

In addition two biochemical features may be mentioned. 1) Chitin is usually absent from deuterostomes, and 2) creatine phosphate is more abundant than arginine phosphate in deuterostomatous phyla, whereas the converse is characteristic of protostomatous groups.

SPERM ENTRANCE

In many species the first observable interaction between the gametes is the bursting of the sperm acrosome when activated by gynogamones (secretion of female gametes, such as fertilizin), or by secretions of cumulus cells in mammals. Lytic enzymes (e.g., hyaluronidase, acid phosphatase, proteases) are released from the acrosome, and help to dissolve egg investments (jelly, vitelline coat, zona pellucida, intercellular cement between follicle cells). An acrosomal tube, in many species—or the side of the sperm head in mammals—can then bind to and fuse with the "egg" cell membrane. Then cataclysmic changes occur in the female germ cell: an action potential appears in the cell membrane owing to changes in its permeability to Na^+ and K^+; calcium ion is redistributed; cortical granules explode; the vitelline coat or zona pellucida elevates (i.e., becomes a fertilization coat), etc.

There are blocking mechanisms to guard against the entrance of more than one sperm (polyspermy). The lifting of a fertilization coat is too slow to be a very effective block. There are species-specific factors which can be extracted from egg membranes that inhibit fertilization. And proteases, released by ruptured cortical granules, can attack sperm-binding macromolecules (glycoproteins). The cortical reaction, however, is relatively slow, appearing several seconds after sperm contact (sea urchin). The fast block to polyspermy is more likely some rapid membrane change, in a fraction of a second, that inactivates the sperm-binding molecules. Large eggs, such as those of birds, commonly exhibit polyspermy, but only one male nucleus usually combines with the female nucleus.

B. Fertilization and Cleavage in Tube Worm

It is highly desirable to supplement the preceding study of fixed material with observations of early development of a living egg. Indeed the use of living material should be encouraged in any course in embryology. Modern technology and rapid transportation make possible the introduction of absorbing exercises in developmental biology in laboratories remote from the seashore or lakes and streams. Aquaria (e.g., *Instant Ocean,* Aquarium Systems, Inc., Wickliffe, Ohio) permit the maintenance of marine organisms in inland schools and air express provides speedy delivery of fresh animals. There is no substitute for actual observation of the union of egg and sperm, the cleavage of a zygote, the formation of a blastopore, the rising of neural folds and their fusion, the rotation of an amphibian embryo within the fertilization membrane owing to ciliary action, the first muscular twitch and beating of the heart, the rocking of a chick embryo within its amnion, and so forth. Motion picture with time-lapse photography is a valuable adjunct to direct observation of living embryos. Many instructive and beautiful films are available nowadays.

A study of fertilization in a polychaete worm, *Mercierella enigmatica,* is very useful. This worm occurs in certain brackish estuarine lakes forming masses of intertwined calcareous tubes on boats and floats. The sexes are separate and eggs and sperm are usually obtainable at any time of the year by removing the worms from their tubes and placing them in clean sea water. The Dahl Company, P. O. Box 566, Berkeley, California 94701, is a source of supply of the animals, via air express.

Each pair of students will be furnished a watch glass and pipette. At the instructor's desk select a small cluster of the tubes and fill the watch glass with sea water. Break the tubes using a pair of forceps until the bodies of the worms are exposed. Grasp a worm with the forceps, gently pull it out of its tube and place it in the watch glass of sea water. Repeat this for two or three additional worms. Shortly the gametes will be shed and will collect about the body of the worm. The eggs, even to the unaided eye, will appear as minute granules whereas the sperm forms a viscid white mass. If both sexes are not present, dissect out other worms until both eggs and sperm are obtained.

Prepare a microscope slide as follows. Smear a *small* amount of vaseline on the heel of the palm of one hand and *gently* draw the edges of a clean cover slip across it to build up a ridge about $\frac{1}{16}$ of an inch high along each edge. Lay the cover slip on the desk with the vaseline upwards and pipette into the center of the cover slip one drop of water containing eggs. To this add another drop of water containing sperm. Only a tiny amount of the sperm suspension is necessary. If the sperm concentration is too great polyspermy and irregular cleavage may result. Finally lower a clean microscope slide onto the cover slip until the vaseline has formed a complete seal. Invert the slide onto the stage of the microscope and observe with intermediate and, with care, high-power magnification. Use 10× oculars. Adjust the illumination so that the lamp is as far from the microscope as the desk top permits in order to reduce the heating of the preparation. Temperatures above 20° C (68° F) will cause abnormal development. Attention should be given to the proper use of the mirror, iris diaphragm, and the condenser of the microscope. Note the following.

Spermatozoa: Observe the swimming motion of the tiny sperm cells. Note the position taken by a sperm when it strikes the surface of an egg. Reflect upon the morphological differences between sperm and egg.

Unfertilized egg (primary oöcyte):

Germinal vesicle: a large irregularly shaped body centrally situated in the cell. The germinal vesicle disappears upon sperm entrance.

Cortical cytoplasm and *vitelline membrane:* a narrow band at the surface of the egg.

Yolk: coarse, refringent granules and spheres in the cytoplasm which impart to the egg a faint yellow color.

Sperm entrance: The entrance of the sperm is rapid and will be observed only by chance. Point of entrance may be identified by the *fertilization cone,* a small elevation on the surface of the egg opposite the head of the sperm. The sperm becomes embedded in the cone and withdrawn into the interior of the egg.

Fertilized egg: The egg into which a sperm has entered may be distinguished from the unfertilized egg by the following features:

1. Disappearance of the germinal vesicle.

2. *Shape:* fertilized eggs are usually spherical or oval whereas unfertilized ones are somewhat irregular in outline. See fig. 10*a, b.*

3. *Polar bodies:* These are tiny cells cut away from the germ cell at the first and second maturation divisions and are designated, respectively first and second polar bodies. The first polar body appears in 20 to 40 minutes after sperm penetration, depending upon temperature, as a small translucent bead. The polar body is formed at a point on the egg known as the *animal pole.* The opposite point on the egg surface is the *vegetative pole;* the imaginary line between, the *animal-vegetative axis.* The lower half of the egg, the *vegetative hemisphere,* is richer in yolk than the *animal hemisphere* and poorer in cytoplasm. These differences are not visible in the tube-worm egg but will be observed in the amphibian egg to be studied later. The second polar body forms 10 to 20 minutes later, typically directly under the first one.

4. *Fertilization membrane:* Almost immediately after sperm penetration a membrane is formed on the surface of the egg and separated from the cortical layer of the ovum by a narrow space, the *perivitelline space.* The membrane and space are seen to good advantage after polar body formation because the bodies elevate the membrane considerably. The vitelline membrane is a part of the fertilization membrane.

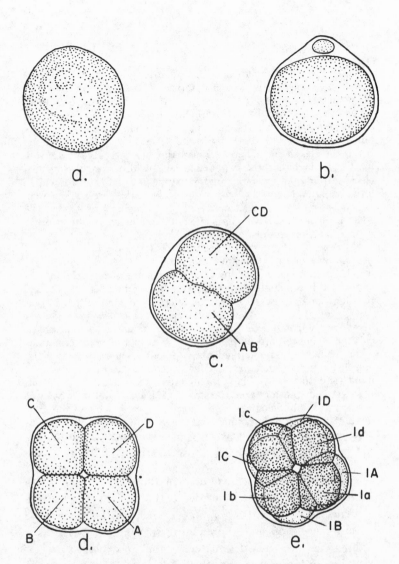

Fig. 10. Early development of the tube worm, *Mercierella enigmatica*. *a*, Unfertilized egg; *b*, fertilized egg showing first polar body; *c*, first cleavage; *d*, second cleavage; *e*, third cleavage. A—B, First four blastomeres; 1 A—1 D, macromeres; 1 a—1 d, micromeres.

Activation is the stimulation of an egg to develop. Usually activation is accomplished by sperm entrance. Some eggs, however, develop naturally without fertilization. Under these circumstances some stimulus other than sperm entrance activates the egg; this is called *parthenogenesis* (G. *parthenos* = maiden + *genesis* = to be born). Among insects one might cite several examples; the drone bee is the classic one. The experimentalist has been able to activate most eggs—including the mammalian egg—with physical, chemical, electrical, or other stimuli. Such experimental activation is called *artificial parthenogenesis*. By careful study of the fertilization process and of parthenogenesis much—although perhaps relatively little —has been learned about activation, this complex physiochemical process which releases the egg to develop.

Fertilization, the predominant form of activation in nature, may best be regarded as a dynamic interaction between egg and sperm which begins even before the two gametes are in physical contact and which is concluded sometime later—many minutes to hours depending upon the species and conditions involved—with the completed zygote endowed with full capacity for development and heredity. The entrance of the sperm sets in motion a chain of physical and chemical processes leading to structural transformations which collectively are termed development or *morphogenesis* (G. *morphe* = form + *genesis*). The egg is emancipated from its moribund state; the events which follow are as remarkable as life itself. A single cell transforms into a new organism through a myriad of biochemical, physiological, and morphological changes, the patterns of which are determined by the hereditary contributions of both egg and sperm but modified by environment. Look at the egg again and ponder on its significance and potentialities!

Cleavages: The segmentation of the zygote begins with the division of the cell into two *blastomeres*. First cleavage occurs about 50 to 60 minutes after sperm entrance. The blastomeres appear to be equal in size. Actually there is a difference; one cell, designated by the embryologist as CD, is slightly larger than the other, AB blastomere. The second cleavage, like the first, is vertical i.e. passes from animal to vegetative pole. The third cleavage is horizontal, or approximately equatorial, giving a quartet of animal blastomeres, called *micromeres* because they are slightly smaller, and a quartet of vegetative blastomeres, called *macromeres*, being larger and containing more yolk. The resulting four cells are designated A, B, C, and D. The study of *cell lineage* is concerned with the identification and fate of the cells established by the several cleavages. According to the scheme of terminology established by early American embryologists, the micromeres are designated, 1a, 1b, etc. and the quartet of macromeres, 1A, 1B, etc. (see fig.10). It is beyond the scope of this study to identify the blastomeres after the third cleavage. One further point, however, should be noted. Observe that the micromeres do not lie directly above the macromeres. The mitotic spindles at the third cleavage were obliquely oriented; as a consequence the micromere and macromere formed from each of the first four blastomeres do not lie one directly over the other. Viewed from the animal pole the micromere (e.g., 1a) lies to the right above the corresponding macromere (1A). To state it differently, the micromeres appear to be rotated slightly to the right or in a clockwise direction so that they lie in the cleavage furrows of the macromeres. The type of cleavage giving this arrangement is called *spiral cleavage*. It is characteristic of annelids and molluscs, for example. In other forms, such as echinoderms and vertebrates, the alignment of the mitotic spindles at third cleavage is not oblique but vertical. Thus, micromeres lie directly above the macromeres. This pattern of cleavage is termed *radial cleavage*. Subsequent cleavages are rapid and difficult to follow. A blastula is formed, which by virtue of cilia, is motile.

C. Early Development in the Mussel

In a hanging-drop preparation, like that described above for *Mercierella,* compare the features of fertilization and cleavage of the mussel egg with those of the tube worm egg.

Polar lobe: a large bulge from the vegetative pole of the egg, just opposite the polar bodies, appearing about twenty minutes after the formation of the second polar body. The cytoplasm of this lobe has been shown by experimentation to be essential for the formation of certain larval structures, an apical organ for example. In the absence of this material the larva is lacking these structures. Eggs which show defective development upon the loss of some part are said to be *mosaic eggs* (those of ctenophores, molluscs, annelids, tunicates, etc.). Eggs of other groups, (e.g., coelenterates, echinoderms, vertebrates, etc.) may develop normally even after the loss of a large part, a half or more, of its cytoplasmic mass. These are said to be *regulative eggs*.

First cleavage: The division of the mussel egg is clearly unequal (see fig. 11). One blastomere, the CD cell, is much larger because the polar lobe is incorporated into it. The other blastomere (AB), lacking polar lobe material, is smaller. If these two blastomeres are separated experimentally, the CD blastomere develops into a normal, although small, larva; the AB blastomere forms a defective embryo similar to that from a lobeless egg.

Cleavage

In miolecithal eggs, such as the starfish egg, there is so little yolk that the mitotic spindle assumes a central position in the zygote and in later blastomeres so that cleavages are both complete (*holoblastic*, G. *holos* = entire or complete + *blastos*) and equal, i.e. the daughter cells resulting from any division are essentially the same size. In medialecithal eggs, such as that of *Mercierella* or *Mytilus*, the third cleavage is above the equator, the spindle being situated nearer the animal pole. The more yolk present, the higher will be the position of the spindle and cleavage plane, and the greater the discrepancy in size between the smaller animal blastomeres and the larger vegetative blastomeres. In megalecithal types, egg of a bird for example, so much yolk is present that cleavage is

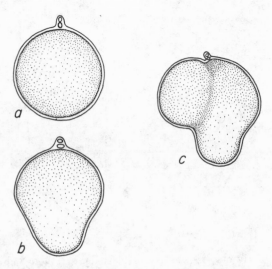

Fig. 11. Early development of the mussel, *Mytilus edulis*. *a*, Formation of polar bodies; *b*, formation of polar lobe; *c*, first cleavage.

confined to a small cap of cytoplasm at the animal pole, the *blastodisc*, and the cleavage planes do not pass through the entire egg but only a short distance. This type of cleavage is termed *meroblastic* (G. *meros* = part + *blastos*) or partial cleavage in contradistinction to holoblastic cleavage.

Whether a cleavage is vertical or horizontal is determined by the orientation of the spindle and this in turn by the longest axis of cell or better, of its cytoplasmic mass, because in cells with large amounts of inert material the long axis of the cell and of its cytoplasmic mass may not be the same. Note that the long axes of the starfish blastomeres in the 4-cell stage are vertical. The spindles for the third cleavage will thus be vertically oriented and the cleavage planes will consequently be horizontal. In a megalecithal egg the blastodisc after the first two cleavages is like a pie with two crossing knife marks on its surface. The four cells, like the thin sectors of a pie, still have their longest axes in a horizontal plane. The third cleavage spindles (see fig. 12) will, therefore, be horizontally oriented and the division planes will be vertical. This difference in position of the third cleavage plane is a basic distinction between holoblastic and meroblastic cleavage patterns.

An experimental test of the relationship between the long axis

51

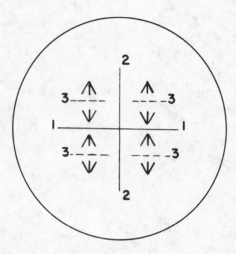

Fig. 12. Early cleavages of avian blastoderm. 1 and 2, First and second cleavage planes; 3, position of third cleavage furrows and the third cleavage spindles.

of the cytosome and the plane of division is provided by subjecting a 4-cell echinoderm embryo to mechanical pressure (e.g. with a cover slip pressing down on the animal ends of the blastomeres) so that the four cells are flattened. The long axes of the compressed cells will now be horizontal and the third cleavages will be, as a consequence, vertical. Incidentally this was one of the early experiments performed in the relatively new science of developmental mechanics or experimental embryology.

The above principles may be expressed in a set of propositions commonly called *rules of cleavage*.

1. The mitotic spindle is centered in the cytoplasmic mass of the cell but may or may not be centered in the cell as a whole depending upon the amount and distribution of yolk or other inert material present.

2. The orientation of a cleavage spindle is determined by the longest axis of the cell's cytoplasmic mass.

3. The plane of cleavage lies at right angles to the spindle and hence to the long axis of the cytoplasmic mass.

4. The rate of cleavage is directly proportional to the concentration of the cytoplasm or inversely related to the concentration of yolk.

Review the material on page 49 regarding the numbering of blastomeres (cell lineage) and the difference between spiral and radial cleavage.

The importance of cleavage: 1) The zygote is "cut up" into smaller units out of which tissues and organs can be formed. 2) Surface area is increased relative to the volume of protoplasm. As all interchanges and interactions between the cells of a living system and between the system and its environment must be effected through cell membranes, surface area is of strategic importance. Included among these cellular interactions are embryonic inductions (see p. 42). 3) In many instances cleavage furthers the segregation of materials of specific morphogenic significance, localized in discrete regions of the zygote, by "packaging" them in particular cells. One is reminded of polar-lobe cytoplasm (see fig. 11) in the zygote of the mussel which at first cleavage is segregated into blastomere CD and at the second cleavage into blastomere D. Or, to cite another example, cleavage parcels the gray crescent of the amphibian zygote among a group of cells that later becomes the dorsal lip region of the gastrula. In the first instance specific "organ-forming materials" (so-called for lack of more exact information about their nature) are segregated into particular cells; in the second example a "material" is distributed to the cells of a particular region which later, because of this "material," will function as the primary organizer of the embryo (see p. 61).

At one time it was thought that cleavage might segregate nuclear material of specific morphogenic significance, but the experiments of Hans Driesch and Hans Spemann showed that the nuclei of early blastomeres are interchangeable. Transplantation of nuclei from embryonic, adult and cultured cells into enucleated eggs has shown that nuclei in different cells from an individual are qualitatively alike, and that all may promote complete and normal development of an egg of that species. Differentiation cannot be due to differences in genic (DNA) information contained within the cells of the embryo. Regulation of transcription of genes is the primary way in which cells become different, basically in the kinds of proteins synthesized. Stated in another way: the supply of mRNA's is the crux of somatic differentiation. So if availability of DNA for transcription is important, there must be regulatory molecules in the cytoplasm. And because the cytoplasms of the cleavage cells differ, so would their nuclear-cytoplasmic interrelations. After each mitotic division the daughter nuclei would be "reprogrammed," depending upon the regulatory molecules in their respective cytoplasms.

D. Early Amphibian Development

Preparations of the early development of a frog or salamander afford a study of the features in the development of a strongly medialecithal egg. Observe the following.

Blastula: a hollow ball of cells, or one-layered embryo, resulting from the segmentation of the zygote; differs grossly from the blastulae just studied in the size and arrangement of the blastomeres.

Micromeres: the smaller cells of the animal hemisphere, forming the roof of the blastocoel. Is this roof a single layer of micromeres? Note that the external surface of the outermost cells shows a thin layer of brown pigment (melanin). The granular character of the cells is due to the presence of yolk platelets. Look for mitotic figures.

Macromeres: the larger cells of the vegetative hemisphere, forming the floor of the blastocoel. Note that the inner cells, which are very large owing to much yolk, are loosely arranged. Do the vegetative cells possess pigment?

Blastocoel: The cavity of the blastula, sometimes called the segmentation cavity, is large and eccentric in position. The granular material observed in the cavity represents the coagulated fluid which filled the blastocoel.

Chorion (Fertilization membrane): the thin membrane surrounding the blastula. This membrane is not secreted by the egg but by the follicle cells in the ovary. Hence, it is a secondary membrane. At the time of fertilization the chorion lifts off of the egg, probably together with an exceedingly fine vitelline membrane—if it exists at all—to form a fluid-filled perivitelline space. The chorion plus vitelline membrane thus constitutes a fertilization membrane. The term chorion unfortunately has another usage, namely, the designation of one of the fetal membranes of amniotes.

54

Gastrula: the two-layered embryo, resulting from inrolling of surface cells of the blastula to form an inner tube or archenteron, which is the primitive gut. Identify the following in a sagittal section of the gastrula.

Ectoderm: the outer layer of cells investing the gastrula.

Archenteron: the inner tube lined with mesentodermal cells. The cavity of the archenteron is designated the gastrocoel.

Mesoderm: here represented by the roof cells of the archenteron. These cells, which are loosely held together, have migrated from the outside of the blastula to take up the position seen here. Note that the surfaces of the cells facing the gastrocoel still carry the pigmented cortical cytoplasm. These cells are for the most part potential notochordal cells and are known to act as the "primary organizer" of the embryo. Note the relationship of the archenteric roof (organizer) to the overlying ectoderm, which is induced by the organizer to form nervous tissue (see p. 61).

Entoderm: represented here by the large, yolk-filled cells forming the thick floor of the archenteron. The sides of the archenteron, not seen in sagittal section, are also entodermal.

Blastopore: the opening of the archenteron to the outside.

Yolk-plug: a mass of large yolky cells which blocks the blastopore so that actually no opening or pore exists. The rounded plug sometimes protrudes from the blastopore.

Dorsal lip: the well-defined upper margin of the blastopore. It is via the upper lip that the greatest inrolling of cells takes place.

Ventral lip: less well-defined lower margin of the blastopore. The ventral lip is marked by a deep cleft.

Topography: The blastopore marks the future posterior end of the embryo; it is not far from the truth to say that the blastopore becomes the future anus (compare with the condition in the echinoderm embryo). The anterior end of the future embryo is thus approximately 180 degrees opposite the blastopore, namely, at the apex of the archenteron. The ectodermal cells above the anterior tip of the archenteron mark the position of the old animal pole. The animal pole has not migrated as it would seem but the entire gastrula has rotated owing to a shift in the center of gravity caused by the forward displacement of the large yolky cells.

Blastocoel: Remnants only may be seen in the gastrula, as for example, the narrow space between archenteric roof and the ectoderm and a somewhat larger but irregular space situated below the antero-ventral wall of the archenteron. This space sometimes contains the coagulated fluid of the blastocoel. Eventually these remnants will be obliterated.

Fertilization membrane: still present as an envelope about the gastrula; note, however, that it is more loosely applied to the gastrula than to the blastula. This is probably owing to a decrease in the diameter of the embryo.

Amphibian Gastrulation

Yolk not only affects the pattern of cleavage but that of gastrulation as well. A miolecithal egg, such as that of a starfish or of amphioxus, gastrulates largely by *invagination,* a process simply defined as an insinking of the vegetal cells. In these forms invagination is presumably due to forces, not yet understood, which cause the vegetal area to buckle into the blastocoel. Change of cell shape is an important feature but whether this is cause or effect is not clear. There are several theories of invagination.

In the amphibian embryo, gastrulation is only initiated by invagination. Moreover, the entire vegetal area does not fold inward, presumably because of the large amount of yolk present. Only certain cells invaginate—those directly below the position of the *gray crescent,* the zone of reduced pigmentation which appeared earlier near the equator of the completed zygote. Change of cell shape is very marked—from more or less spherical to elongated, bottlelike form with the narrow neck of each cell attached to the forming blastoporal groove and the large bulblike body of the cell protruding into the blastocoel.

As soon as the blastoporal groove is established the cells above it (dorsal lip) undergo mass movement into the groove which sinks deeper into the embryo. The dorsal lip is thus rolled into the interior of the young gastrula. This inrolling is termed *involution.* It is the predominant gastrular process in the amphibian. The blastoporal groove becomes crescentic and its lateral lips likewise involute. Eventually the blastopore becomes circular and inrolling occurs at the ventral lip also. As cells roll inside their places are taken by cells moving toward the lips. This blastoporeward movement is called *epiboly* (G. *epi + ballein* = to throw). Actually epiboly and involution are but segments of one general movement toward and through the blastopore into the interior.

The extent and character of these gastrular movements have been clearly shown by following regions of the late blastula stained with vital dyes, such as neutral red and nile blue sulfate. By many careful mapping studies it has been possible to project on to the late blastula or beginning gastrula the pattern of organ-forming areas. This pattern is called a *fate map* (see fig. 13). All of the regions below the line A-B are involuted through the blastopore leaving on the exterior of the completed gastrula only the two ectodermal districts: prospective neural plate and prospective epidermis.

Of the mesodermal districts those which will form head mesoderm and notochord involute over the dorsal lip; somites, lateral plate, and tail mesoderm over the lateral and dorsolateral lips; and entoderm over the ventral and ventrolateral lips. It will be observed that a small amount of future entoderm precedes head mesoderm on the dorsal lip of the blastopore. Cells of this area will line the anterior wall of the archenteron.

The future germ layers—ectoderm, entoderm, and mesoderm—become discrete layers in the course of gastrulation as the districts below line A-B are rolled inside. The ectoderm is simply the presumptive neural plate and epidermis which remain outside to form the superficial layer of the completed gastrula. The other two layers become differentiated by slow separation along the plane (line as seen in the fate map) between future lateral plate and entoderm. As these districts involute over the ventrolateral lips they part and

57

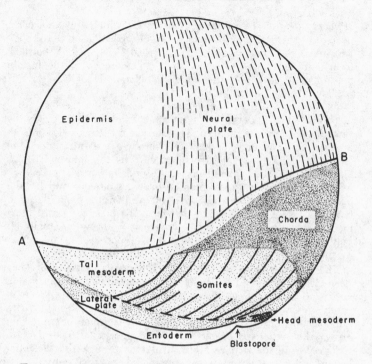

Fig. 13. Amphibian fate map (urodele). (After Vogt, 1926.)

slip by one another, the lateral plate moving laterally and ventrally, the entoderm being carried medially and dorsally. The end result is essentially two cups: (1) an inverted mesodermal cup composed of future notochord medially, somitic material dorsolaterally, and lateral plate on the sides; (2) an entodermal cup, or archenteron, nestled inside the outer mesodermal one. The sides of the entodermal cup soon extend dorsally beneath the notochord, fuse with each other, and thereby establish an entodermal roof to the archenteron. Contrary to older views it is inadvisable to hold that mesoderm of the amphibian embryo develops from either ectoderm or entoderm. Forerunners of all three exist as areas of the blastula, morphologically undefined but nevertheless real as demonstrated by vital staining experiments, and come into existence as discrete layers in the course of gastrulation as described above.

Neurula (G. *neuron* = nerve): the embryo in the process of forming the neural tube or primordium of the central nervous system. Study the cross sections of the neurula and identify the following structures.

Ectoderm: the outer layer investing the neurula; may be subdivided now into two regions.

Neural (medullary) plate: a dorsal, thickened plate of ectodermal cells. In the early neurula the neural plate is relatively flat and bounded on each side by a low ridge, the neural fold. In older neurulae, the plate becomes U-shaped and the neural folds are quite high. In the late neurula the neural plate has become folded into a tube, the neural tube, by the fusion in the midline of the neural folds. Its cavity, the neurocoel (G. *neuron* + *koilos*) is thus a space cut off from the outside. Note that the cells of the plate are columnar, that they contain fine melanin granules at their outer ends, and that their nuclei have become elongated. Do the cells contain yolk platelets?

Presumptive epidermis: the remainder of the ectoderm which will form the outer layer of the integument. Compare these ectodermal cells with the neuroepithelial cells just described.

Mesoderm: The archenteric roof of the gastrula has now separated from the entodermal sides and floor of the primitive gut and broadened to form a distinct intermediate germ layer. In the early neurula this is a flat layer beneath the neural plate with curved lateral wings extending down on each side of the embryo between the ectoderm and the yolky entoderm. In the late neurula the mesoderm may be differentiated as follows.

Notochord (G. *noton* = back + L. *chorda* = cord): rounded mass of mesodermal cells lying in the midline

59

directly beneath the neural plate. In longitudinal aspect the notochord would be seen as a strand of cells along the main axis of the body.

Somites (G. *soma* = body) or *epimere* (G. *epi* + *meros*): blocks of mesoderm, somewhat pyramidal in shape, flanking the notochord on each side and lying beneath the lateral parts of the neural plate.

Lateral plates or *hypomere* (G. *hypo* = under + *meros*): the remainder of the mesoderm seen here or, as described above, the wings which grow ventrally between the ectoderm and entoderm. Eventually these down growths will meet and fuse at the mid-ventral line. The demarcation between somite and lateral plate is not yet clear.

Nephrotome° (G. *nephros* = kidney + *tomos* = a cutting or part) or *mesomere:* a narrow region of the mesoderm between the somite and lateral plate not yet differentiated; also called intermediate mesoderm.

Coelom: In some older neurulae a narrow cleft may be seen forming in the lateral plate separating it into an outer layer, somatic mesoderm, and an inner layer, splanchnic mesoderm (G. *splanchnon* = an entrail). The coelom of the amphibian, as well as of other vertebrates, thus arises from a splitting of the mesoderm and is known as a schizocoel (G. *schizein* = to split + *koilos*). In amphioxus and other protochordates, which are assumed to be ancestral to the vertebrates, part or all of the coelom is enterocoelic in origin. Perhaps primitive chordates possessed an enterocoel but in the vertebrates the coelom has become secondarily a schizocoel.

Entoderm: the large cells, heavily laden with yolk, surrounding the gastrocoel and constituting the definite archenteron. The secondary roof, of entodermal cells, is just being formed by the fusion in the midline of the lateral archenteric walls.

Fertilization membrane: still present as a loose envelope about the neurula. The presumptive epidermis will soon develop cilia by means of which the embryo slowly rotates within the membrane.

DIFFERENTIATION AND INDUCTION

The central feature of morphogenesis is differentiation, which means development from the relatively simple to the complex, from the relatively homogeneous to the heterogeneous—literally becoming different. Out of the zygote, which incidentally is far from simple and homogeneous, come cells, tissues, and organs of exceedingly complex and varied organization. This differentiation frequently involves the action of one part of the developing system upon another part, a process called embryonic induction (see p. 42). The part from which the influence emanates is the inductor or organizer or evocator; the part affected is the reaction system. As a result of induction the fate of the latter is irreversibly determined: that is, the kind of cell or tissue which it will form is irrevocably fixed.

The classic example of embryonic induction and the one most studied is the formation of nervous tissue from simple ectoderm by the organizing action of chordamesoderm or its precursor the dorsal lip of the blastopore (Hensen's node of higher vertebrates). As neural evocation involves the establishment of the axial organs of the body it is called primary induction. Since the discovery of this phenomenon by the German embryologist Hans Spemann in the early 1920's other examples of organizers have been revealed. Most of them are instances of localized—hence secondary—inductions as, for example, the evocation of a lens from epidermal ectoderm by the optic vesicle. Moreover, a great effort by many workers throughout the world has been made in the last fifty-five years to elucidate the mechanism of induction. Although much has been learned the precise nature of both the stimulus and the response remains largely unknown. The following tentative conclusions appear to be justified at this time.

1. Induction is an important and widespread process, especially among vertebrates, providing an ordered and harmonious differentiation of a large embryonic system.

2. Inductions probably result from several physico-chemical reactions involving inductive agents. One current theory holds that in primary induction there are at least two principles: neuralizing and mesodermalizing factors.

61

The evidence for this theory is based largely upon inductions in amphibian gastrular ectoderm by killed tissues from various adult animals. Examples: killed guinea pig liver induces anterior head (archencephalic) structures such as large brain vesicles, nasal organs, eyes, and balancers; dead kidney from an adder evocates posterior head (deuterencephalic) structures such as hindbrain and ear vesicles; spinocaudal structures differentiate in response to killed guinea pig kidney; and lastly, alcohol-treated bone marrow of the guinea pig induces mesodermal structures (e.g., somites and nephric tubules).

3. The neuralizing factor may be a ribonucleoprotein, the mesodermalizing factor a protein unassociated with RNA. The former appears to be thermostable and soluble in organic solvents; the latter is highly heat labile and insoluble in some organic solvents. The active principle of the neuralizing factor is probably not in the nucleic acid part of the macromolecule, but in the protein moiety. Tiedemann has characterized the mesodermal factor as follows: soluble in phenol without irreversible denaturation; inactivated by proteolytic enzymes and by SH-reagents such as mercaptoethanol; and molecular weight of about 28,000. The highly purified protein induces entodermal structures (pharyngeal and intestinal epithelium) as well as mesodermal tissues. It is as yet uncertain whether there is one substance or two closely related factors.

4. It has been postulated that gradients of neuralizing and mesodermalizing substances in the embryo result in regional aspects of differentiation. The gradient of the neuralizing substance extends from high anteriorly and dorsally to low posteriorly and ventrally; that of the mesodermalizing factor is from high posteriorly to low anteriorly with a secondary gradient, as in the neuralizing factor, from high dorsally to low ventrally. Regional development results, according to this theory, from interaction of the two gradients along both the anterior-posterior axis and along the dorso-ventral axis.

5. Under certain experimental conditions the inducing stimulus is effective across short intercellular distances as shown by the passage of inductive substances through millipore filters 60-80 microns in thickness with pores 0.4 micron in diameter. Processes of the inducing cells extend into the pores, but only about 8 microns. Filters with pores 0.1 micron in diameter permitted induction only when the filters were no thicker than 20 microns.

6. Inductions may be evoked after relatively short exposure to the inductor (e.g., three or four hours).

7. The competence or sensitivity of the reaction system to the inductor may be temporally or spatially limited. Examples: the

competence of ectoderm to form nervous tissue is restricted temporally to the period of gastrulation; the competence of epidermal ectoderm to differentiate lens may be spatially restricted in certain amphibians to the head or even to the presumptive lens ectoderm itself.

8. The species characteristic nature of the induced structure rests with the reaction system, not with the inductor.

Just how does an organizer or evocator or inductor (these words have slightly different connotations) act? Do they impart specific "information" to the reaction system; do they introduce some specific chemical substance; or is their action strictly non-specific, just a triggering stimulus? As of January, 1971, the picture is far from clear. It has been established, however, that in some instances (e.g., differentiation of neurones and pigment cells) that all the "information" and machinery and chemical substrates and enzymes are present within the competent ectoderm (presumptive epidermis). The same may be true for other cell types. All that is needed to induce differentiation is some sort of "shock," as simple as an ionic shift (e.g., change in pH). It is hard to reconcile this interpretation, however, with the differences in differentiation which result from the use of different kinds of inductors (e.g., neuralizing versus mesodermalizing inductors).

Modern molecular biology, based largely upon genetic and biochemical studies of microörganisms, offers a possible explanation. The theory of Jacob and Monod for control of enzyme production in bacteria might be extended to the phenomenon of induction as follows. The differentiation of a particular structure (e.g., lens) may require the syntheses of specific proteins which in turn are dependent upon certain genetic units, called operons in the Jacob-Monod theory. Operons for the several lens proteins appear to be inactive until the early tailbud stage of development owing, perhaps, to inhibition by repressor substances (produced by regulator genes in the Jacob-Monod theory). Perhaps the lens-inducing substance produced by the optic vesicle functions as a de-repressor, when it enters presumptive lens cells at the critical moment in development, by combining with repressor substances and thus releasing the operons from inhibition.

A new dimension in research on inductors has been opened by Lester and Lucena Barth in recent years (see Barth and Barth, *Developmental Biology*, 20:236-262, 1969). They have shown that induction of neurones and pigment cells is dependent upon the concentration of sodium ion. The inductive effect of cations (lithium, calcium, magnesium, etc.) occurs at a sodium chloride concentration of 0.088M but not at 0.044M. Similarly, *normal induction* of neurones and chromatophores in small explants of lateral ectoderm to which are added small explants of dorsal lip is dependent on the external concentration of sodium. The sodium dependency is restricted to the period of normal induction. The working hypothesis of the Barths is: "inducing compounds, both natural and unnatural, have as a common factor an alteration in cell membrane properties resulting first in release and redistribution of inorganic ions to new binding sites."

The use of nucleopore filters (possessing smooth, straight pores which millipore filters lack) and scanning electron miscroscopy have shown, in the instance of corneal induction by lens capsule, that processes of the epithelial cells pass through the filter (pore size as small as 0.1 μm) to contact the extracellular matrix of the lens thereby inducing corneal differentiation (collagen synthesis and stroma formation). These studies emphasize the importance of the processes of the reacting cells penetrating the transfilter and of extracellular substances from the inductor (e.g., collagen and glycosaminoglycans in the instance of the lens). Finnish workers, using the same methods to study neural induction in amphibians, reached similar conclusions. Initial triggering of neural differentiation results from macromolecules from the organizer entering the pores of the filter to meet processes of ectodermal cells. But, regionalization of the neural plate requires contact between ectodermal processes and mesoderm.

Review Questions

1. Distinguish between activation and fertilization. How can an egg be artificially activated (artificial parthenogenesis)?

2. Name some morphological and physiological changes occurring in the egg at the time of sperm entrance.

3. What is the function of the fertilization coat?

4. State the rules of cleavage.

5. How could one test experimentally the third rule?

6. Does cleavage segregate nuclear materials for special purposes? Cytoplasmic materials?

7. What is the gray crescent? What is its significance?

8. Compare the third cleavage plane in the starfish (or amphioxus), the amphibian, the bird, and the placental mammal.

9. Compare gastrulation in starfish, frog, bird, and rat.

10. Name derivatives of somite, lateral plate, nephrotome.

III

DEVELOPMENT IN
THE 10 MM. FROG LARVA

Serial cross sections of the 10 mm. tadpole of the Leopard Frog, *Rana pipicns,* present a good introduction to vertebrate organogenesis, the development of organs. A few regions of the body have been selected for study. Before examining the first level a brief review of the major divisions of the nervous system and of planes of symmetry should be helpful.

VERTEBRATE NERVOUS SYSTEM

The vertebrate nervous system may be subdivided as follows:
  A. Central nervous system (CNS) = brain and spinal cord.
  B. Peripheral nervous system = cranial and spinal nerves and ganglia or, in other words, all neural elements outside the CNS.
  1. Cerebro-spinal system (CSS) = that part of the peripheral system carrying nerve fibers primarily to skeletal muscle (somatic motor neurones) and from somatic sense organs (somatic sensory neurones).
  2. Autonomic nervous system (ANS) = that part of the peripheral system carrying visceral motor fibers (to smooth muscles and glands) or visceral sensory fibers (from internal sensory endings). In places autonomic neurones run in the CSS as, for example, in the dorsal and ventral roots of spinal nerves. The ANS is further subdivided into sympathetic and parasympathetic systems. Review the distinctions between these subsystems.

In both embryology and anatomy the CNS is a convenient reference because of its segmentation (metamerism). Very soon after neurulation the neural tube becomes subdivided into brain vesicles and the spinal cord. The embryonic brain vesicles are at first three in number, then five as listed below with their respective cavities (subunits of the neurocoel) and adult derivatives (see p. 67).

(see p. 67)

PLANES OF SYMMETRY

The plane of symmetry of bilaterally symmetrical systems is the *sagittal plane* (L. *sagitta* = arrow), the only one which bisects the system into equivalent halves. A plane is defined by the two axes through which it passes. The sagittal plane is that passing through the antero-posterior axis (the chief axis) and the dorso-ventral axis. A *transverse plane* (cross-sectional), such as those to be studied now, is determined by the dorso-ventral axis and the right-left axis. A *frontal plane* passes through antero-posterior and right-left axes. It is sometimes useful to refer to a *parasagittal plane* (G. *para* = along side of) which is one to either right or left of the sagittal plane. Hence, a parasagittal section would not pass through the antero-posterior axis and accordingly not bisect the body into equal halves.

A. Level of the Telencephalon and Nasal Organs

Examine quickly the first few sections of the 10 mm. frog larva and then select those at about the level of figure 14 for careful study. Identify the following structures.

Telencephalon (G. *tele* = far + *enkephalos* = brain): the anterior part of the embryonic forebrain, the prosencephalon (G. *pros* = toward or front + *enkephalos*), already subdivided sagittally into two large bodies, roughly hemispheric, flattened against each other in the midline. Each contains a cavity, a *lateral ventricle*, formed by an evagination of the side of the neural tube at the anterior end of the neurocoel.

Layers of brain wall:

Ependymal layer: innermost layer, one-cell thick lining of neurocoel. The cells send extensions to the outer surface of the neural wall and become supportive in function.

66

Divisions of Vertebrate Brain

Primary vesicles	Secondary vesicles	Cavities	Adult derivatives
Prosencephalon (forebrain)	Telencephalon (divided into right and left vesicles)	Lateral ventricles (Ventricles I and II)	Cerebrum
	Diencephalon	III ventricle*	Epithalamus, thalamus, hypothalamus
Mesencephalon (midbrain)	Mesencephalon	Cerebral aqueduct	Optic lobes, etc.
Rhombencephalon (hindbrain)	Metencephalon	IV ventricle	Cerebellum
	Myelencephalon	IV ventricle	Medulla oblongata

* A part of the third ventricle extends into the telencephalon in the midline.

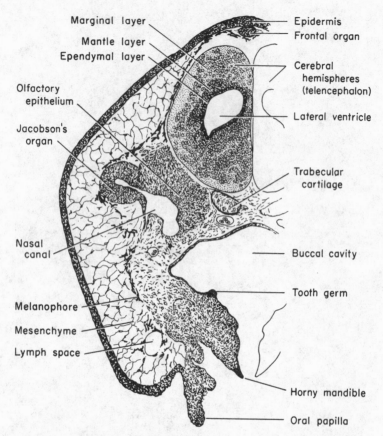

Fig. 14. Cross section of 10 mm. frog larva at the level of telen-cephalon and nasal organs.

Internally they possess cilia which move the cerebral spinal fluid in the neurocoel.

Mantle layer: broad layer adjacent to the ependymal layer; later becomes the gray matter of the CNS.

Marginal layer: outermost fibrous layer containing scattered nuclei; later becomes the white matter of the CNS.

Nasal organ: thick-walled tube lying ventrolateral to the telencephalic hemispheres; formed by invagination of the ectoderm. The medial lining of the nasal canal will become the *olfactory epithelium* from which the *olfactory nerve* will extend to the olfactory lobes of the brain. Trace the nasal cavity forward to its entrance from the outside. The opening is the *external naris* and marks the point of the original ectodermal invagination. Now trace the passageway to the buccal cavity. This opening is the *internal naris* or *choana.*

Frontal organ: a small medial body situated beneath the epidermis and above the telencephalon. This body arises by evagination from the roof of the diencephalon together with the *epiphysis* (see fig. 15) and "migrates" forward to the position seen here. Young specimens will show the outgrowth as yet undivided into epiphysis and frontal organ. Recent electron microscopy has revealed that the frontal organ is a "third eye" as it possesses typical vertebrate photoreceptors.

Vomeronasal (Jacobson's) organ: a lateral sacculation of the nasal organ. It is believed to function in picking up olfactory sensations from food in the buccal cavity.

Buccal cavity: spacious chamber, lined with a flat epithelium, into which mouth and nasal passageways open; derived from the stomodeum. What germ layer lines this cavity? If the tadpole is sufficiently old the jaws will be tipped with a brown horny (cornified) material. External to the jaws are lobose *oral papillae. Tooth germs* may be seen on the inside of the lower jaws.

Melanophores (G. *melano* = black + *phoros* = carrying): light brown stellate cells over the dorso-lateral surface of the brain and lateral to the nasal organs. They are magnificently shown in the first sections of the animal. Examine them under high magnification; note fine granules of melanin (light brown individually; black in masses). The melanin can be moved in and out of the processes of the cell

69

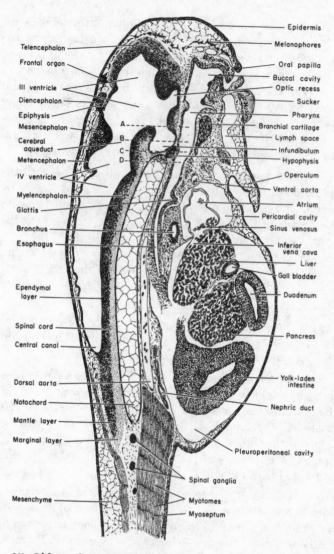

Telencephalon

Frontal organ

III ventricle

Diencephalon

Epiphysis

Mesencephalon

Cerebral
aqueduct

Metencephalon

IV ventricle

Myelencephalon

Glottis

Bronchus

Esophagus

Ependymal
layer

Spinal cord

Central canal

Dorsal aorta

Notochord

Mantle layer

Marginal layer

Mesenchyme

Epidermis

Melanophores

Oral papilla

Buccal cavity

Optic recess

Sucker

Pharynx

Branchial cartilage

Lymph space

Infundibulum

Hypophysis

Operculum

Ventral aorta

Atrium

Pericardial cavity

Sinus venosus

Inferior
vena cava

Liver

Gall bladder

Duodenum

Pancreas

Yolk-laden
intestine

Nephric duct

Pleuroperitoneal cavity

Spinal ganglia

Myotomes

Myoseptum

Fig. 15. Oblique longitudinal section of 10 mm. frog larva. Dorsal part of head and trunk shown in sagittal plane; ventral half of body and tail cut parasagittally.

(how controlled?), thus effecting changes in the color of the animal.

Cartilages: In older specimens small masses of hyaline cartilage (*prechordal* or *trabecular cartilages*) may be found beneath the telencephalic hemispheres. These will later become part of the cartilagenous cranium (*chondrocranium*). Examine the tissue under high power. Identify clear *matrix, lacunae,* and *chondrocytes* (G. *chondro* = cartilage + *kytos* = hollow vessel, hence cell).

Mesenchyme: The spaces between the epidermis and the above organs are filled with mesenchyme, a loose reticulum of mesodermal cells which are stellate in appearance and migratory in behavior. The outermost of these cells will form the inner layer of the integument or *dermis*.

Epidermis (G. *epi* = upon + *derma* = skin): the outer two-cell thick layer of the skin derived from ectoderm. Examine under high power. It contains much free melanin; a few *epidermal melanophores* may be seen.

B. Level of the Diencephalon and the Eye

Diencephalon: median vertically elongated brain vesicle. Its cavity is *III ventricle* (see fig. 16).

Infundibulum (L. funnel): Trace the diencephalon posteriorly. Note that it seems to be constricted frontally into a dorsal tube and a smaller ventral component with thin roof and thick sides. The latter is the infundibulum, a ventroposterior evagination of the diencephalon. The cavity of the infundibulum is III ventricle. Identify cross sections corresponding to levels A and B of the sagittal section shown in figure 15.

Mesencephalon: the part of the brain dorsal to the infundibulum. Its cavity is the *cerebral aqueduct*. The sub-

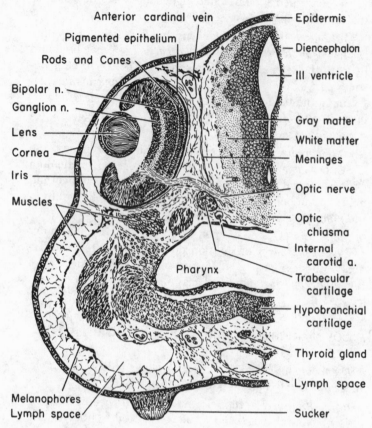

Anterior cardinal vein
Pigmented epithelium
Rods and Cones
Bipolar n.
Ganglion n.
Lens
Cornea
Iris
Muscles

Epidermis
Diencephalon
III ventricle
Gray matter
White matter
Meninges
Optic nerve
Optic chiasma
Internal carotid a.
Trabecular cartilage
Hypobranchial cartilage
Thyroid gland
Lymph space
Sucker

Pharynx

Melanophores
Lymph space

Fig. 16. Cross section of 10 mm. frog larva at the level of diencephalon and eyes.

division of the midbrain into the future optic lobes is foreshadowed by a median dorsal notch.

Pituitary body: Trace the infundibulum caudally. Note that an oval mass, the pituitary, appears below its thin floor. The pituitary is also called hypophysis (G. *hypophyein* = to grow beneath). Compare with level C of figure 15. It is difficult at this time to identify the individual lobes of the pituitary (pars glandularis, pars intermedia, and pars nervosa). Continue tracing posteriorly. The hypophysis disappears and the tip of the *notochord,* flanked by *parachordal cartilages,* comes into view. Compare with level D of figure 15.

Eye: Is there an embryological basis for the poetic expression, "The eye is the window of the mind"?

Retina: the thick inner layer of the optic cup. The following sublayers show early differentiation.

> *Ganglion cells:* innermost layer of cells. Axons of these cells turn back through the retina as a cable and run from the eye to the floor of the diencephalon (*optic chiasma*) as the *optic nerve.* Identify the nerve if the specimen is sufficiently old. Is this a true nerve? Why are these neurons called ganglion cells?

> *Rod* and *cone cells:* the outermost layer of cells. From the outer end of each cell (adjacent to pigmented layer) a photoreceptoral process (rod or cone) is forming. With what layer of the brain wall is the layer of receptor cells homologous?

> *Bipolar neurones:* the intermediate layer of cells which will synapse with receptor and ganglion cells. What is a bipolar neurone? Unipolar? Multipolar? Name examples.

Pigmented epithelium: the thin, one-cell thick, pigmented outer layer of the optic cup. Function? What forms the *iris* of the eye?

Lens: spherical body partly enclosed by the optic cup. What is its origin?

> *Lens epithelium:* the one-cell thick outer layer.

> *Lens fibers:* the columnar cells forming the core of the lens; later they become long narrow transparent fibers arranged in layers.

73

Cornea: the superficial covering of the eye composed of an outer layer of ectoderm plus mesodermal elements beginning to assemble between the ectoderm and lens.

Choroid and *sclera:* the outer investments of the optic cup represented at this time by mesodermal cells aggregating outside the pigmented epithelium.

Pharynx: flat but broad gut at this level. What germ layer lines it?

Cartilage: In addition to small blocks of cartilage below the diencephalon, parts of the chondrocranium, several masses or long bars of *hypobranchial cartilage* will be seen under the floor of the foregut. These are parts of the *visceral skeleton* which supports the pharynx.

Thyroid: a pair of small bodies beneath the hypobranchial cartilage. From what germ layer do they come?

Skeletal muscle: In older specimens mesodermal masses lateral and ventral to the pharynx may be identified as incipient branchial muscles. Examine under high power. Are cross-striations visible?

Suckers or *adhesive disks:* a pair of hillocks on the ventral surface of the tadpole consisting of columnar cells. Examine under high power. What is the position of the nuclei? What is the material in the distal parts of the cells? Function of the organ?

C. Level of the Myelencephalon and Ear

Myelencephalon: large brain vesicle with thick floor (*basal plates*) and thin roof, the latter to become the vascularized *posterior choroid plexus.* The cavity is *IV ventricle.* See figure 17.

Ear (auditory vesicle): irregular hollow organ to each side of the medulla.

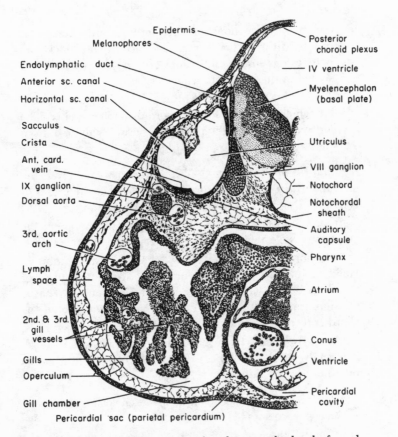

Fig. 17. Cross section of 10 mm. frog larva at the level of myelencephalon and ears.

Endolymphatic duct: thick-walled tube lying against
the medulla; it marks the course of the invagination of
the auditory vesicle from the ectoderm.

Utriculus: the large dorsal chamber of the auditory
vesicle not yet well delimited.

Semi-circular canals: anterior, horizontal and posterior sacculations of the utriculus. In older specimens

the two connections of the anterior and horizontal canals to the utriculus may be traced. Note thickened patches of sensory epithelium (*cristae*), especially well shown in the horizontal canal.

Sacculus: ill-defined, ventral part of the auditory vesicle which will later form a more distinct chamber. The *lagena* or *cochlea*, the organ of hearing, will arise from the sacculus.

Auditory (acoustic) ganglion: the dense mass on the medial surface of the auditory vesicle. What is the origin of ganglia? How does the auditory (VIII) nerve form?

Auditory capsule: condensation of mesenchyme about the auditory vesicle which will form a cartilagenous enclosure.

Notochord: sagittal skeletal rod beneath the hindbrain. How far forward does it extend? Describe the features of the chordal cells as seen under high magnification. What is the significance of the vacuoles? The mesenchyme about the chord is forming the *notochordal sheath.*

Heart: By tracing from section to section work out the relationships of the chambers of the heart, a slightly coiled tube bent to the right.

Pericardial cavity: coelomic space enclosing the heart. What is a coelom?

Conus: the anteriormost chamber, connected to the *ventral aorta* (truncus), a short median vessel outside the pericardial cavity. Note paired *aortic arches* (see below) emerging from the ventral aorta. Trace the conus (also called bulbus) as it swings to the right (see fig. 17).

Ventricle: the next chamber, moving caudally, joined to the conus on the right and ventrally (at about x in fig. 17). The ventricle has exceptionally thick and spongy walls. Are you looking at the anterior surface of the section shown in figure 17? Does the ventricle become subdivided into two chambers?

Atrium: large thin-walled, dorsally situated chamber. In older specimens the atrium may show beginning subdivision into right and left chambers.

Sinus venosus: posteriormost chamber; lies mostly to the right and at the cranial end of the liver. Note the entrance of the *hepatic vein (inferior vena cava)* from the liver and the *comon cardinal veins* from the lateral body walls.

Gill chamber (opercular cavity): large paired chambers, continuous with the cavity of the gut, one on each side of the heart. Within the chambers note several folded *internal gills* bearing branchial blood vessels. What has happened to the external gills of the early larva? Trace the course of waterflow through the animal. Give an account of external respiration.

Dorsal aortae: paired vessels, one above each gill chamber.

Aortic arches: embryonic vessels encircling the pharynx, now modified in relation to the gills. If time permits, attempt to trace one of the branchial vessels from the ventral aorta, through the gill, and to the dorsal aorta. The aortic arches involved are numbers 3-6 (the first two aortic arches—mandibular and hyoid—being transitory) because visceral arches 3-6 are gill bearing. What is a visceral arch?

Miscellaneous structures: Facial (VII) ganglion: large mass of nerve cell bodies anterior to but fused with the auditory ganglion (the two together called *acoustico-facialis*); *Trigeminal (V) ganglion:* a still larger mass immediately anterior and dorsal to the acoustico-facialis ganglion. *Metencephalon:* the anterior part of the hindbrain, not yet well defined, behind the optic lobes and medial to V ganglion. *Glossopharyngeal (IX) ganglion:* dark body below each auditory vesicle. *Pericardial sac* or *parietal pericardium:* wall of pericardial cavity. *Visceral pericardium (epicardium):* the surface epithelium of the heart itself. *Operculum:* the external wall of the opercular cavity formed by a downgrowth or fold of body wall which fuses with the body posteriorly creating a spacious gill chamber that remains open to the exterior on the left side via an opercular pore. Compare with the atrium of amphioxus. Are the internal gills really inside the animal?

D. Level of First Spinal Nerves and Pronephros

Spinal cord: The section shown in figure 18 is at the approximate junction of medulla and spinal cord, although some posterior choroid plexus may still be seen. The cavity of the spinal cord is the central canal, a continuation of the IV ventricle. A *fiber tract* (tractus solitarius?) consisting of light nerve fibers may be observed embedded within the mantle layer. This is a good example of a descending (or ascending) bundle of axons. Note a beginning formation of the *meninges,* the covering, of the CNS. What is the difference between cord and chord?

First spinal ganglia: aggregations of nerve cell bodies at the ventro-lateral surface of the cord. What kind of neurones are represented by these cell bodies?

Myotomes (G. *mys* = muscle + *toma* = section or part): segmental blocks of skeletal muscle flanking the notochord. Examine under high power. The muscle fibers, now differentiating, are longitudinally arranged, hence seen in cross section.

Pleuroperitoneal cavity: the large coelomic cavity in which all viscera, except the heart, are suspended. Why is this compound name used? Is a viscus really inside the coelom?

Esophagus: a tube with folded entodermal lining (differentiating into the mucosa) situated below the notochord.

Dorsal aortae: paired blood vessels between gut and chorda. Trace them posteriorly to note their fusion into a single aorta.

Stomach: Trace the esophagus into the large ventrally

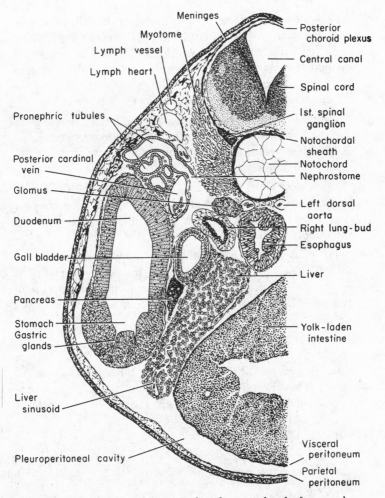

Fig. 18. Cross section of 10 mm. frog larva at level of pronephros.

placed, thick-wall stomach. Observe deep evaginations of the entodermal lining to form the rudiments of the *gastric glands*.

Duodenum: Trace the gut dorsally to the duodenum in the upper right corner of the body cavity.

Yolk-laden intestine: Further tracing of the gut is not very useful. Examine under high power, however, a part which is packed with pale oval *yolk platelets*. These are packages of food from the egg which were parcelled out to the cells by mitosis. The yolk may have disappeared in most organs by this time, depending upon the age of your specimen.

Liver: an organ to the right of the midline of loose organization of *liver cords* interspersed with *sinusoids*. What is the difference between a sinusoid and a capillary? Observe associated *gall bladder* (see fig. 18).

Bile duct: Trace posteriorly until the gall bladder disappears. Shortly in the same area a thick-walled tube, the bile duct, appears within the liver. Trace it to its entrance into the duodenum.

Pancreas: To the right of the liver and bile duct (posterior to the level of fig. 18) a large organ, the pancreas, makes its appearance. Note that its organization differs from that of the liver, having *alveoli* or nests of cells around very small ducts.

Lung-buds: thick-walled ovals, one on each side of the esophagus. You may see yolk platelets in the lining cells of these structures. Why? Trace the lung-buds anteriorly in the series. They become thin-walled *bronchi* which are connected medially to the *trachea* which opens in turn into the gut via a slit, the *glottis*.

Pronephros (G. *pro* = forward + *nephros* = kidney): paired larval excretory organs in the dorso-lateral angles of the body cavity. From what segment of the mesoderm do they arise?

Pronephric tubules: tortuous ducts of cuboidal epithelium, three in number, at somite levels 2-4. Observe that the tubules are supplied by the *posterior cardinal veins*.

Nephrostome (G. *nephros* + *stoma* = mouth): the opening of each tubule into the coelom.

Nephric duct: Proceed caudally in the series to the end of the mass of nephric tubules. One duct only remains; this is the nephric duct into which the tubules have opened. Trace the duct posteriorly. It moves medially. Note its partner on the opposite side. Continue to trace the pair in large jumps (by rows of sections) until they join and empty into the *cloaca*, the terminal segment of the gut.

Glomus: a highly vascularized body lateral to the dorsal aorta.

Review Questions

1. What is the central nervous system? Name the divisions of the brain, their respective cavities, and their derivatives in the adult.

2. Name in order and in detail the layers of cells pierced by a theoretical probe passing into the 10 mm. frog larva along the right-left axis through the pupil of the eye and into the neurocoel.

3. Name in order and in detail the parts of the circulatory and excretory systems of a 10 mm. frog larva which transports a molecule of urea from its site of formation in the liver to the pond water outside the animal.

4. What is the embryonic source (i.e., the immediate developmental precursor) of the following: semicircular canals, buccal cavity, nephric duct, lens, Jacobson's organ, frontal organ, epiphysis, infundibulum, non-neural part of hypophysis, rod and cone cells, lung-buds, olfactory lobes, optic lobes, endolymphatic ducts.

5. From what germ layer do the following come? utriculus, gall bladder, gastric glands, visceral peritoneum, meninges, pronephric tubules, thyroid gland, cornea.

6. Define: glomus, pleuroperitoneal cavity, visceral peri-

toneum, nephrotome, cerebral aqueduct, auditory capsule, parachordal cartilages, chondrocranium, nerve tract, nerve, meninges, cerebro-spinal system, parasympathetic system, nerve cell body, sinusoid, epithelium.

Derivatives of Neural Crest

The neural crests are "jacks of many trades." What a system for study by the experimental embryologist and developmental geneticist! A partial list of the various types of structures which differentiate from these aggregations of ectodermal cells follows.

1. Cerebro-spinal ganglia. Clusters of crest cells remaining relatively close to the neural tube differentiate into sensory neurones and constitute the cranial and spinal ganglia. Not all neurones in cranial ganglia are of crest origin, however; many arise from ectodermal placodes, especially exteroceptive neurones. What is an exteroceptive neurone? A proprioceptive neurone?

2. Sheath (Schwann) cells. These elements are highly migratory at first but they settle down on naked nerve fibers (axis cylinders) that grow out from neuroblasts within the neural tube or in aggregations of crest cells (ganglia). Each sheath cell encloses a segment of the fiber forming its outer investment or neurilemma. In myelinated (medullated) fibers the sheath cell becomes wrapped about the axis cylinder in such a fashion that layers of its plasma membrane are added to form a thick insulating coat called the myelin sheath.

3. Autonomic ganglia. These colonies of neurones are derived from migratory neuroblasts of crest origin. It is noteworthy that their axons are non-medullated, i.e., the sheath cells in which they are embedded do not lay down concentric lamina of their cell membranes.

4. Meninges. The membranous coverings of the brain and spinal cord.

5. Pigment cells. Various chromatophores (color cells), such as melanophores (black cells), erythrophores (red cells), guanophores (silvery, reflecting cells), etc. arise from migratory crest cells.

6. Adrenal medulla. The inner component of the adrenal gland is formed by migratory crest cells, and also other chromaffin bodies, so-called because they stain with chromates, indicating the presence of amines such as epinephrine.

7. Islet of Langerhans (pancreatic islets). The islets differentiate into several cell types, the most important being beta cells (produce insulin) and alpha cells (secrete glucagon). What is the action of glucagon?

82

IV

DEVELOPMENT OF THE CHICK EMBRYO

A. Development before Incubation

Fertilization occurs as soon as the egg enters the oviduct of the hen; and as the egg passes down the oviduct and uterus, the tertiary membranes (structures formed outside of the ovary such as dense and thin albumen, inner and outer shell membranes, and calcareous shell) are added and development proceeds through cleavage and into gastrulation. Cleavage, which is meroblastic, is at first confined to a small cap of cytoplasm, known as the *blastodisc* (G. *blastos* + *diskos* =platter), situated at the animal pole and resting on the large mass of yolk (See p. 52). The result of cleavage is not a ball of cells as in holoblastic types, but a plate of cells termed the *blastoderm* (G. *blasto* + *derma*). Between the blastoderm and the yolk is a narrow cavity, the blastocoel. Gastrulation, briefly, consists of two phases: 1) hypoblast formation by delamination (L. *de* = down + *lamina* = plate) from the blastoderm, and 2) mesoderm formation. The first phase only occurs before laying and this to different degrees in different eggs, depending upon whether the egg is laid in the late afternoon or retained by the hen overnight. This variation in the extent of hypoblast formation at the time of laying accounts in large measure for differences in the ages of incubated eggs. Another possible factor might be variation in temperature of the incubator. Usually incubators are regulated for 100°–101° F (37°–38° C).

B. Development in the 15–25 Hour Chick Embryo

1. WHOLE MOUNT OF THE 15–16 HOUR CHICK EMBRYO: Examine under low and intermediate power only. Owing to the thickness of whole mounts high-power objectives can not be used. Breakages can be largely prevented by attention to this point and to proper procedure for focusing, namely, lowering the objective to a position close to the cover slip, *watching from the side*, and then with eye at the ocular bringing the preparation into focus by raising the barrel of the microscope. Identify the following structures.

Blastoderm: a double-layered disc of cells, which has been lifted from the yolk, stained, and mounted on the slide. In general, the blastoderm may be divided into two regions.

Area pellucida (L. *area; lucidus* = clear): the lighter, centrally situated, pear-shaped region. The lightness of the area pellucida is owing to the thinness of the blastoderm (see fig. 19) which consists of an outer layer of epiblast (G. *epi = upon + blastos*), from which ectoderm, mesoderm, and embryonic entoderm will differentiate, and an inner layer of hypoblast (G. *hypo* = under + *blastos*) (see fig. 19 and p. 86). With the formation of a layer of hypoblast, the cavity beneath might be called gastrocoel instead of blastocoel (see fig. 19).

Area opaca (L. *area; opacus* = dark): the darker region surrounding the area pellucida. The density of the area opaca is due to yolk clinging to its under surface. Two subregions of the area opaca may be listed, although not identified in the preparation.

Zone of junction:° the inner and broader region of the area opaca in which the cells are not cut off from the yolk (see fig. 19). The zone of junction moves pe-

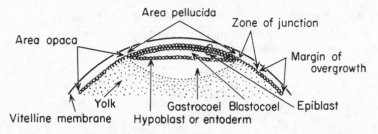

Fig. 19. Schematic cross section of the chick blastoderm prior to formation of primitive streak.

ripherally, and cells cut free from the yolk are added to the area pellucida.

*Margin of overgrowth:** the outer and narrower region of the area opaca in which the cells are growing peripherally and freely over the yolk.

Primitive streak: slightly thicker (hence darker) streak lying in the main axis of the area pellucida.

Primitive groove: slight depression or trough extending along the middle of the primitive streak; appears as a light stripe because of its thin floor.

Primitive ridges or *folds:* low ridges in the blastoderm, one on each side of the primitive groove; being thicker they will appear as darker streaks. Primitive groove and primitive ridges together constitute the primitive streak.

Topography:

Posterior end: identified by the end of the primitive streak nearer the zone of junction (i.e., boundary between areas pellucida and opaca); also identified by the narrower or "neck" end of the usually pear-shaped area pellucida.

Anterior end: identified by the end of the streak farther from the zone of junction or by the broad end of the area pellucida.

85

Primitive pit: the anterior end of the primitive groove where the blastoderm is more deeply indented. The primitive pit does not always show well.

Primitive knot or *Hensen's node* (after Victor Hensen, German physiologist, 1835–1924): small, darker region immediately anterior to the primitive pit; not always clearly shown.

AVIAN FATE MAP

A map of prospective organ-forming areas (fig. 20) has been constructed for the chick embryo with techniques (vital staining, carbon marking, radioactive tracers) similar to those used in making the amphibian fate map (see p. 56 and fig. 13). The avian map shows the disposition of these areas, such as neural plate, notochord, somites, etc., on the blastoderm at the close of the first phase of gastrulation or hypoblast formation. Comparing figures 13 and 20 one can see a fundamental similarity in the topographic arrangement of the various districts of the amphibian and avian embryos. The essential difference is that in the amphibian they lie on the surface of a ball (blastula) whereas in the bird they are on the surface of a disc (blastoderm). The similarity of relationships may be illustrated by noting that the order of districts in the sagittal plane of the amphibian blastula and chick blastoderm is the same: head mesoderm, notochord, neural plate, and epidermis. As in the amphibian, so in the bird, all areas below the line A-B will be moved beneath the surface—in the frog and salamander through a blastopore, which is at first a groove, then a crescent, then a large circle and finally a small circle; in the bird through the primitive streak which is the homologue of the amphibian blastopore. The anterior end of the streak, or primitive pit, is homologous with the frog's early blastoporal groove or, to state it differently, with the middorsal sector of the amphibian blastopore.

The formation of hypoblast (first phase of gastrulation) has been a subject of controversy. The generally-held opinion is formation by delamination—that is, separation of cells from the under surface of the blastoderm. This process begins near the posterior border of the blastoderm and proceeds anteriorly some hours before laying of the "egg" by the hen. At first the hypoblast is an incomplete layer of cells under the epiblast; later, after incubation is begun, the hypoblast is established in the entire area pellucida. The hypoblast will

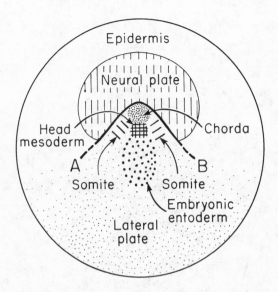

Fig. 20. Avian fate map before formation of primitive streak. (After various sources.)

later assume an extraembryonic position, as explained below. Hypoblast, or extraembryonic entoderm, is important in at least two respects: it will be the absorptive layer of the yolk-sac (a structure described later) that encloses the yolk; and it contains the primordial germ cells that later come to lie in a crescentic region of the blastoderm anterior to the embryo (see p. 18).

The second phase of gastrulation consists of the involution of embryonic entoderm and mesoderm from the epiblast through the primitive streak (see below). Newer studies have demonstrated by radioactive tracers that future *embryonic* entoderm arises not from hypoblast but from a district in the midline of the epiblast (heavily stippled region shown in Fig. 20). Small transplants, labelled with tritiated thymidine, were followed during gastrulation to establish this point. As the embryonic entoderm moves through the primitive streak, which forms first within that district, the hypoblast is shifted to an extraembryonic position. Mesodermal districts then involute through the primitive streak.

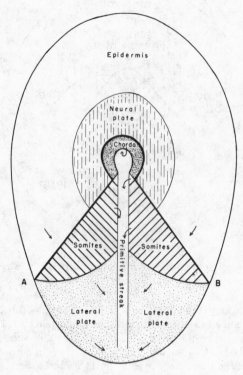

Fig. 21. Avian fate map at primitive streak formation. (After Pasteels, 1940.)

Figure 21 shows the positions of the districts of the avian embryo in the second phase of gastrulation in which mesoderm is formed. Head mesoderm has already involuted through the primitive pit and is no longer visible externally. Notochord is now being involuted (shown by arrow) over the dorsal lip. The long sides or ridges of the streak are homologous with the lateral lips of the amphibian blastopore. Over these primitive ridges move lateral plate and somites (see arrows). Since lateral plate rolls in first it comes to lie farther laterally in the completed gastrula, somitic material involuting in the path of lateral plate takes up a more medial position. Again, as in the amphibian embryo, the wheeling surface movement of the districts (see arrows) may be termed *epiboly*. When the mesodermal regions have completely involuted through the streak only the ectodermal areas of neural plate and epidermis remain outside. The mesodermal districts are disposed inside as a layer between the ectoderm and entoderm (see fig. 22). In the floor of the primitive streak, however, the identity of the germ layers is not clear; this is a characteristic feature of the streak. Until the mesodermal districts have completely moved into the interior it is incorrect to refer to the upper layer of the blastoderm as ectoderm. Figure 22 is too diagrammatic. The mesodermal cells creep along, by means of pseudopodia, on a basal lamina covering the under surface of the ectoderm, as shown by scanning electron microscopy. The basal lamina is secreted by the ectoderm in advance of the migrating front of mesoderm.

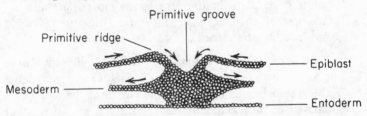

Fig. 22. Schematic cross section of the chick blastoderm showing formation of primitive streak. Arrows show movement of cells to establish the mesoderm.

2. WHOLE MOUNT OF THE 18–19 HOUR CHICK EMBRYO: Examine under low and intermediate power only. Identify the following structures.

Primitive streak, groove, ridges, pit and *knot:* essentially as described above for the 15–16 hour chick embryo.

Head process (old-fashioned name for the notochord): faint streak anterior to the primitive pit, resulting from the involution of presumptive notochord (Hensen's node) through the primitive pit and forward between the ectoderm and entoderm.

Mesoderm: In the whole mount the regions of the blastoderm which contain mesoderm are slightly darker, owing to the fact that they are thicker than those composed of only two layers.

89

Anterior border of the mesoderm: indicated by a faint line extending across the area pellucida at the level of the anterior end of the notochord (head process). Mesoderm moving through the primitive pit has migrated forward to this level. Note that the blastoderm is darker (thicker) posterior to this line.

Lateral border of the mesoderm:° faint line in the area opaca roughly parallel to the boundary between area pellucida and area opaca; not clearly shown, except in excellent preparations. This line means that mesoderm, involuting via the primitive streak, has moved laterally between ectoderm and entoderm beyond the area pellucida and into the area opaca.

Proamnion (G. *pro* = before + *amnos* = literally lamb, but referring to membrane about the foetus): a clear region of the area pellucida anterior to the border of the mesoderm composed of only ectoderm and entoderm; mesoderm will not migrate into this region until much later in development.

Neural plate: Although not visible in the whole mount, the ectoderm above the notochord is thickening into a plate which will soon roll up into the neural tube.

3. WHOLE MOUNT OF THE 20–21 HOUR CHICK EMBRYO: Examine under low and intermediate power. Identify the following structures.

Neural folds: two short, thick streaks situated close to one another at the level of the anterior border of the mesoderm. These represent the folded edges of neural plate which will soon meet and fuse in the mid-line and thereby form the neural tube.

Neural groove: the clear streak between the neural folds.

Head fold: usually a crescentic dark line just anterior to the neural folds (see fig. 23, *a*) indicating that the blastoderm is folded along this line. This fold which becomes more prominent and soon more U-shaped is the future head

of the embryo. At first only the ectoderm is folded but soon the entoderm likewise and the head fold exhibits two crescentic lines (see fig. 23, *b*), the more anterior being ectoderm, the more posterior, entoderm. Another crescentic line, if present, joining the ends of the inverted U-shaped head fold (see fig. 23, *c*) represents the posterior boundary of the fold which is undercutting the head.

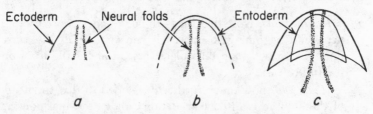

Fig. 23. Stages in formation of head fold.

Notochord: now a discrete structure, indicated by a faint longitudinal band extending caudad from the region of the neural folds into Hensen's node. It is bounded on each side by a narrow white streak. By careful focusing explore the dorsoventral relations of crests of the neural folds, floor of the neural groove, and notochord.

First intersomitic grooves: The first (and in some specimens the second also) intersomitic grooves will be seen as paired, transverse, clear streaks just above Hensen's node and lateral to the notochord. The small mesodermal blocks between the first and second intersomitic grooves are designated somite no. 2. Somite no. 1, anterior to the first intersomitic groove, is continuous with the mesoderm of the future head. The number of somites present may be used very roughly to estimate the age of the embryo in terms of hours of incubation by adding the number of somites to 20. Thus, an embryo with 3 somites would be approximately 23 hours old.

Lateral plate: This region of the mesoderm can be identified as the thinner (lighter) mesoderm lateral to the thicker

mesodermal somites and unsegmented mesoderm of the future head.

Other structures previously studied: proamnion, anterior border of the mesoderm, primitive streak and its parts, Hensen's node, and areas of the blastoderm.

4. WHOLE MOUNT OF THE 23–25 HOUR CHICK EMBRYO: Observe under low and intermediate power only.

Neural folds: Note that they are now longer, more prominent, and beginning to approach one another in the midline. Fusion will soon take place, first at about the level of the future ear.

Head fold: now much further developed so that the head of the embryo is lifted considerably above the blastoderm. Determine this by careful focusing. The outermost line, still roughly an inverted U, may now be designated as the ectoderm of the head. Note that it is continuous with the neural folds.

Foregut: a dark dome-shaped structure lying inside the head with fairly sharp lateral walls of entoderm which lie inside and parallel to the ectoderm of the head. The foregut, although very short now, is the forerunner of the anterior regions of the digestive tube.

Anterior intestinal portal: the posterior boundary of the foregut, represented by the prominent arched line (entoderm) beneath the neural folds. Note by very careful focusing that this line is continuous with the lateral walls of the foregut. The anterior intestinal portal is actually the opening into the foregut (see fig. 24) and the arched line referred to above is the posterior limit of the floor of the foregut, or the folded entoderm at the point X on the diagram.

Subcephalic pocket: the ectodermal lined space beneath the head of the embryo. The posterior wall of the pocket or the folded ectoderm at the point Z in figure 24 may be seen as a faint arched line just cephalic to the anterior intestinal

portal. Note that this line (ectoderm) is continuous with the ectodermal sides of the head.

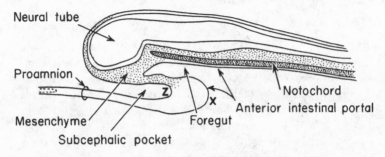

Fig. 24. Sagittal section through 25-27 hour chick embryo.

Mesenchyme of the head: a rather broad stripe of cells lying between the foregut and the ectoderm of the head. Mesenchyme may be regarded as the embryonic connective tissue.

Mesoderm: note the following features.

Anterior border: Locate the anterior border of the mesoderm, situated similarly to that in the 21 hour chick, to the side of the head; trace it laterally on the blastoderm and then medially into the embryo as a fine line just caudal to the line representing the posterior wall of the subcephalic pocket. Not all specimens will show this well.

Thickened splanchnic mesoderm: a heavy line, immediately lateral to each side of the anterior intestinal portal, caused by the thickening of the splanchnic mesoderm, the first step in the formation of the heart. Observe that this line is continuous with the anterior border of the mesoderm. These features will be better understood in the study of the cross sections.

Notochord: now a considerably elongated strand. How far does it extend anteriorly? posteriorly?

Somites: Count them. Somite no. 1 is becoming more distinct.

Unsegmented mesoderm: the stripe posterior to the somitic region and immediately lateral to the notochord, which will later be cut into somites.

Area opaca: now divided into two subregions.

Area (opaca) vasculosa: an inner zone of the area opaca which has many splotches giving the area a mottled appearance. The patches are clusters of cells, known as *blood islands,* which are differentiating into blood vessels and corpuscles. The vessels being outside the embryo belong to the extra-embryonic circulatory system. The peripheral border of the area vasculosa represents the extent of migration of the mesoderm laterally.

Area (opaca) vitellina: the outer zone of the area opaca into which mesoderm has not yet migrated.

Other structures previously studied: area pellucida, proamnion, Hensen's node, and primitive streak. Has the primitive streak changed in length?

5. TRANSVERSE SECTIONS OF THE 24 HOUR CHICK EMBRYO: Begin the study of the series with the anteriormost sections through the head of the embryo and proceed caudally. Correlate cross sections and the whole mount just studied. After identifying a structure under low or intermediate power, study under high power its histological and cytological detail.

a. Sections at the level of the head of the embryo.

Ectoderm (embryonic epidermis): layer of cuboidal cells enclosing the head.

Neural folds: thick ridges of neuroepithelium (neural ectoderm) composed of columnar pseudostratified cells. Have the folds fused at any point? If so, where?

Neural groove (or future *neurocoel*): the depression or incipient cavity formed by the rising neural folds. The floor and sides of the groove are the neural plate now being converted into the neural tube.

Foregut: a dorsoventrally flattened tube ventral to the

neural plate; lined with a layer of squamous entodermal cells.

Oral plate: Note that some sections show the floor of the foregut almost fused with the ventral ectoderm of the head. Here both entoderm and ectoderm are thickened. This plate of entoderm and ectoderm is termed the oral plate or pharyngeal membrane which later breaks to establish the oral opening or mouth.

Mesenchyme: loosely scattered stellate cells in the space about the foregut and developing neural tube.

Notochord: a small medial mass of cells between the foregut and the neural plate. Does the notochord extend anteriorly to the tip of the head?

Proamnion: the region of the blastoderm under the head of the embryo, made up of two germ layers, ectoderm (nearer the embryo) and entoderm.

Subcephalic pocket: the space between the proamnion and the head of the embryo. Trace posteriorly until the ectoderm of the head becomes fused to the ectoderm of the blastoderm. Note that the posterior tip of the subcephalic pocket is now lined on all sides with ectoderm. Trace farther. What becomes of this pocket? Relate these sections to the point Z in figure 24.

Mesoderm: the intermediate germ layer found between the ectoderm and entoderm of the blastoderm lateral to the proamnion. Note that the mesoderm is split horizontally into two layers.

Somatic mesoderm: the upper layer adjacent to the ectoderm; somatic mesoderm and ectoderm together are termed *somatopleure* (G. *soma* + *pleura* = side).

Splanchnic mesoderm: the lower layer adjacent to the entoderm; splanchnic mesoderm and entoderm together are termed *splanchnopleure* (G. *splanchnon* + *pleura*).

Coelom: the cavity between somatic and splanchnic

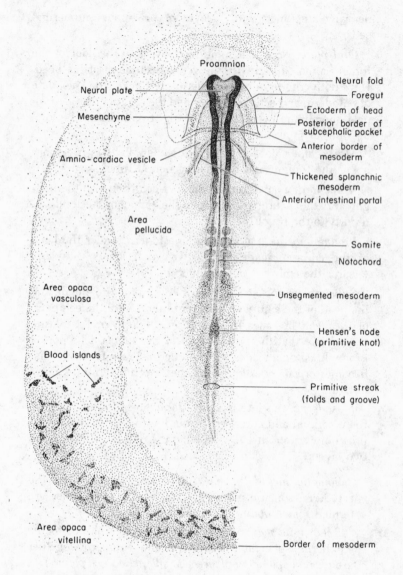

Fig. 25. Whole mount of 24 hour chick embryo.

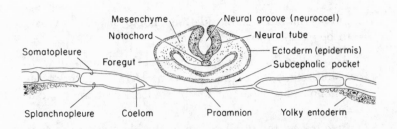

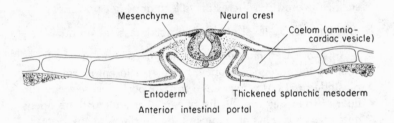

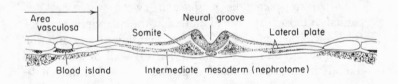

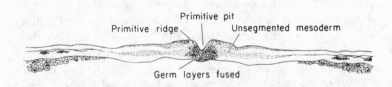

Figs. 26-29. Cross sections of 24 hour chick embryo. Fig. 26. At level of head; Fig. 27. At level of anterior intestinal portal; Fig. 28. At a level posterior to the portal; Fig. 29. Through the primitive pit.

mesoderm. Situated outside the embryo this cavity is said to be extra-embryonic coelom.

Area pellucida: the thinner region of the blastoderm without yolk clinging to the entoderm. Here it includes the proamnion plus a part of the blastoderm containing mesoderm.

Area opaca: the distal region of the blastoderm in which the entoderm is thick owing to yolk adhering to its under surface. Note the spheres of yolk under high power.

b. Sections at the level of the anterior intestinal portal.

Anterior intestinal portal: Locate the section in which the floor of the foregut becomes fused to the entoderm of the blastoderm. In the next one or two posterior sections the floor of the foregut will "disappear." The foregut here opens onto the yolk. This opening is the anterior intestinal portal. Relate these sections to the point X in fig. 24.

Thickened splanchnic mesoderm: Two or three sections caudal to those just studied, note that the mesoderm is quite thick on either side of the lateral margins of the anterior intestinal portal. This thickened vertical wall of splanchnic mesoderm was responsible for the line observed in the whole mount (see page 93).

Amnio-cardiac vesicle: The thickened splanchnic mesoderm identified above forms the proximal wall of the large amnio-cardiac vesicle, so-called because it is involved in the formation of the amnion and the heart. The amnion develops later from its dorsal wall, and the heart is beginning to form now from its thickened proximal wall. The dorsal wall of the vesicle is somatopleure, its ventral wall splanchnopleure. Its cavity is coelom.

Neural crests: At this level the neural folds are fusing in the midline. Note that at the point of fusion the edges of the neural plate are being brought together to complete the neural tube and that above, the epidermis (ectoderm) is

98

likewise fusing in the midline. Between the neural tube and the epidermis, on each side, is a compact mass of cells known as neural crests. Observe under high power. From these cells several structures are derived, such as ganglia, pigment cells, adrenal medulla, sheath cells, etc.

Midgut: the region of the gut posterior to the anterior intestinal portal which possesses no cellular floor.

c. Sections at the level of the somites. Examine the structures studied above and note in addition:

Somites (epimere): three-sided blocks of mesoderm ventrolateral to the neural plate. Compare sections passing through a pair of somites with those passing between pairs of somites.

Nephrotome (mesomere or *intermediate mesoderm):* an ill-defined region of mesoderm immediately lateral to the somites consisting of a narrow unsegmented plate of loosely arranged cells.

Lateral plate (hypomere): the mesoderm distal to the nephrotome. Is it subdivided here into somatic and splanchnic layers?

d. Sections at the level of the primitive streak. Proceed caudally in the series of cross sections from the somitic region of the embryo through the region of the segmental mesoderm and into the primitive streak. The chief feature to be noted now is the fusion in the midline of the three germ layers. Identify:

Primitive ridges and *primitive groove.*

Primitive pit: a depression, slightly deeper and more V-shaped than the primitive groove, situated at the anterior end of the primitive streak.

Hensen's node: the thickened, slightly depressed plate of fused germ layers immediately anterior to the primitive pit.

Area vasculosa: the inner region of the area opaca in which, in favorable sections, may be seen small dark masses of mesodermal cells, the *blood islands.*

Review Questions

1. Draw a sectional view through a hen's egg enclosed in its membranes. Label shell, inner and outer shell membranes, air chamber, thin and dense albumen, chalaza, vitelline membrane, blastodisc, white and yellow yolk.

2. What is the function of the yolk, albumen, and shell? How is embryonic respiration accomplished?

3. Compare amphibian and bird with respect to: cleavage, entoderm formation, blastopore, and organizer.

4. Sketch a sagittal section of the 24 hour chick embryo (A-B in fig. 30) and several cross-sectional views of the same embryo (C-D, E-F, G-H, and I-J in fig. 30).

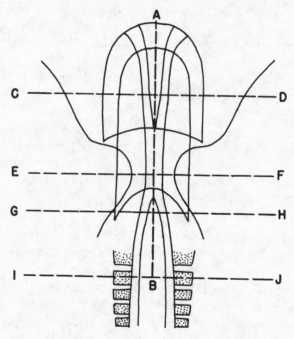

Fig. 30. Twenty-four hour chick embryo.

5. Sketch the fate map for the early chick blastoderm before the formation of the primitive streak. Show by arrows the surface movements on the epiblast as the mesoderm is being formed. Compare with the amphibian fate map.

6. How is the subcephalic pocket formed? the anterior intestinal portal?

7. Distinguish between a section through the neural groove and one through the primitive groove.

8. To what extent has mesoderm migrated in the blastoderm of the 24 hour chick embryo?

9. Define: blood islands, splanchnopleure, head process, nephrotome, area opaca vitellina, proamnion, amnio-cardiac vesicles, oral plate or pharyngeal membrane, mesenchyme.

SECONDARY ORGANIZERS

Spemann (see back cover of manual) called the dorsal lip of the blastopore, or archenteric roof, the primary organizer because it determines the symmetry and axiation of the body, the central nervous system, sense organs, and other structures. He designated other inducers, acting later in development, as secondary organizers. The following are some examples of secondary organizers: the optic vesicle induces a lens; the lens, a cornea; the auditory vesicle, a cartilagenous otic capsule; the infundibulum, a pars intermedia of the pituitary; the subdivisions of the ureteric bud, metanephrogenic nephrons (capsule with glomerulus and tubule); ventral entoderm, a heart; the mesenchyme of glands (e.g., salivary gland), secretory lobules and acini; the ectodermal ridge of the tip of a limb-bud, a foreleg or wing or hind limb. Note that in the last example the inducer is the ectoderm, the reacting cells are mesodermal.

The organizer principle can be extended to the development of hair, feathers, scales, and teeth in which there are inductive interrelationships between ectodermal and mesodermal components. Even organelles can be organizers. The axial centriole induces a cilium when it lies under a cell membrane. The nine doublets of microtubules in the ciliary axoneme are extensions of two tubules in each centriolar triplet. Why does not the third centriolar microtubule "grow out" into the cilium? And what organizes the central singlets of the axoneme? Unanswerable questions at present.

C. Development in the 32–38 Hour Chick Embryo

1. WHOLE MOUNT OF THE 32–38 HOUR CHICK EMBRYO: Examine under low and intermediate power only. If owing to age or staining of your specimen a particular feature can not be favorably observed, borrowing of slides is permissible providing transfer of slides from one student to another is done carefully and with consideration.

Two major differences between the 24 and 33 hour chick embryos are immediately apparent: 1. the differentiation of the neural tube into the major divisions of the central nervous system (see p. 65), and 2. the differentiation of the heart. The first contractions of the heart commence at about 29 or 30 hours of incubation (9-somite stage). The first beats are in the ventricular myocardium; the pulses are slow and irregular. A few hours later the rate increases and the atrium becomes the pacemaker. Still later the sinus venosus takes over this function, and the rate is further increased. Meanwhile blood corpuscles are forming in the blood islands (see below) in the blastoderm. At about 38 hours of incubation intraembryonic and extraembryonic vascular channels become connected with each other, and the corpuscles begin to circulate in a jerky stream, owing to the absence of valves in the system. Identify the following structures.

Prosencephalon (G. *pros* = toward + *enkephalos* – brain): the forebrain; the anteriormost division of the neural tube, consisting of a median vesicle, plus a lateral outpocketing on each side.

Optic vesicles: the lateral evaginations just mentioned.

Anterior neuropore: a median notch or cleft at the anterior tip of the neural tube marking a region where the neural folds have not yet fused.

Infundibulum (L. *infundibulum* = funnel): a ventral outpocketing of the prosencephalon represented by a

102

heavy concave line continuous with the caudal border of the optic vesicles.

Prosocoel (G. *pros* + *koilos*): the cavity of the prosencephalon which later becomes ventricles I to III. The cavities of the optic vesicles—known as opticoels—are parts of the prosocoel.

Mesencephalon (G. *mesos* + *enkephalos*): the midbrain or the division of the neural tube posterior to the prosencephalon, appearing as an oval vesicle, the cavity of which is the *mesocoel* or the future *cerebral aqueduct* or *aqueduct of Sylvius* (after François Sylvius, French anatomist, 1614–1672).

Rhombencephalon (G. *rhombos* = spinning top, hence, having rhomboid shape + *enkephalos*): the hindbrain; the last of the three primary vesicles from which the brain develops. The cavity of the hindbrain is ventricle IV and the rhombencephalon is divided into two regions.

Metencephalon (G. *meta* = between + *enkephalos*): the smaller, almost spherical vesicle caudal to the mesencephalon, the cavity of which is the *metacoel* (a part of ventricle IV).

Myelencephalon (G. *myelos* = marrow + *enkephalos*): a series of four small subdivisions of the neural tube— known as *neuromeres* (G. *neuron* + *meros*)—lying between the metencephalon and the spinal cord. The cavity of the myelencephalon is the *myelocoel* (the remainder of the fourth ventricle). It may be mentioned here that the divisions of the brain supposedly represent two or more neuromeres fused together, as follows: prosencephalon 3; mesencephalon 2; metencephalon 2; myelencephalon 4. Total neuromeres for the brain is 11.

Spinal cord: the neural tube posterior to the myelencephalon; bulges and irregularities in the spinal cord are not considered neuromeres.

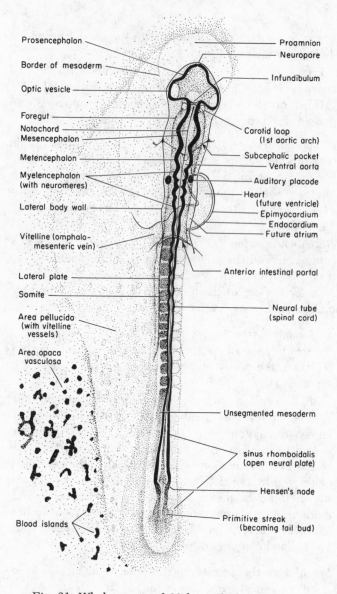

Prosencephalon

Border of mesoderm

Optic vesicle

Foregut
Notochord
Mesencephalon

Metencephalon

Myelencephalon
(with neuromeres)

Lateral body wall

Vitelline (omphalo-
mesenteric vein)

Lateral plate

Somite

Area pellucida
(with vitelline
vessels)

Area opaca
vasculosa

Blood islands

Proamnion
Neuropore

Infundibulum

Carotid loop
(1st aortic arch)

Subcephalic pocket
Ventral aorta

Auditory placode

Heart
(future ventricle)
Epimyocardium
Endocardium
Future atrium

Anterior intestinal portal

Neural tube
(spinal cord)

Unsegmented mesoderm

sinus rhomboidalis
(open neural plate)

Hensen's node

Primitive streak
(becoming tail bud)

Fig. 31. Whole mount of 33 hour chick embryo.

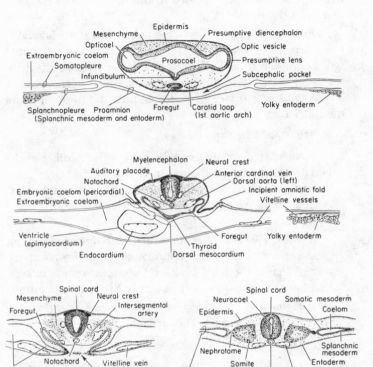

Figs. 32-35. Cross sections of 33 hour chick embryo. Fig. 32. At level of optic vesicles; Fig. 33. At level of ventricle of heart; Fig. 34. At level of anterior intestinal portal; Fig. 35. At the level of a somite.

Sinus rhomboidalis (L. *sinus* = curve, hence, concavity + G. *rhombos*): a region near the posterior end of the neural tube where the unfused neural folds enclose Hensen's node and the primitive pit.

Auditory placodes (G. *plakos* = flat plate): In favorable older preparations a dark spot (thickening in the ectoderm) will be observed on each side of the neural tube at about the level of the second neuromere of the myelencephalon. These are the auditory placodes or pits, the forerunners of the inner ears.

Mesodermal structures:

Notochord: a strand extending along the midline of the body from the infundibulum to Hensen's node. In the region of the brain the notochord is narrow, dark, and wavy; in the trunk region it becomes broader, lighter, and straighter.

Somites: How many somites have differentiated in your preparation? What is the approximate age of your specimen? What is the level of the anteriormost somite in relation to the neural tube? in relation to the intestinal portal?

Unsegmented mesoderm, nephrotome, and *lateral plates:* as described for the 23–25 hour embryo.

Entodermal structures:

Anterior intestinal portal: a distinct arched line at about the level of the first or second somite in the 14–15 somite embryo. Explain the posterior "migration" of the anterior intestinal portal. Focus carefully to determine whether the line (entoderm) lies above or below the neural tube. Is the embryo mounted dorsal side up?

Foregut: Carefully trace faint lines (entoderm) extending forward from the margins of the anterior intestinal portal. These mark the sides of the foregut. The anterior tip of the foregut is usually obscured by the overlying neural tube.

Circulatory system: Study the features described below on the whole mount and then review the system in the demonstration preparation of a 36 hour chick embryo injected with india ink.

Heart: a vesicle or tube, lying cranial to the anterior intestinal portal, with a pronounced bend to the right. The terms right and left are used with reference to the embryo and not the observer. A good ventral view of the heart may be obtained by carefully inverting the slide and supporting it by two coins placed appropriately on the stage of the microscope.

Epimyocardium (G. *epi* = upon + *myos* = muscle + *kardia* = heart): the heavy line which outlines the heart. It represents a layer of mesodermal cells which will later form the epithelial covering of the heart (epicardium or visceral pericardium) and the cardiac musculature (myocardium).

Endocardium (G. *endon* = within + *kardia*): a thin line, inside the epimyocardium, forming an inner tube. It represents a layer of mesodermal cells which will later form the inner epithelial lining of the heart.

Prospective chambers of the heart: The heart at this stage is roughly a U-shaped tube, with the base of the U directed to the right, showing faint outlines of the future chambers (see fig. 36).

Atrium (L. *artium* = hall): the posterior arm of the U-shaped heart; the forerunner of the auricles.

Ventricle (L. *ventriculus* = stomach, or meaning cavity of an organ): the base of the U.

Conus (L. *conus* = cone): the anterior (smaller) arm of the U.

Sinus venosus (L. *sinus; vena* = vein): the region of the heart situated immediately cranial to the anterior intestinal portal.

107

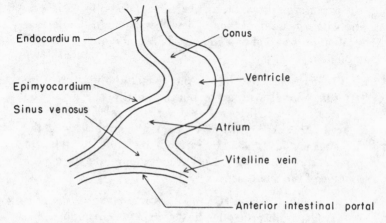

Endocardium

Conus

Epimyocardium

Ventricle

Sinus venosus

Atrium

Vitelline vein

Anterior intestinal portal

Fig. 36. Dorsal view of heart in 33 hour chick embryo.

Vitelline veins or *omphalomesenteric veins* (G. *ompha-los* = navel + *mesaraic* = mesentery): two large vessels emerging from a plexus of vessels on the blastoderm to course medially just cranial to the lateral borders of the anterior intestinal portal and to empty into the sinus venosus of the heart (see fig. 36).

Ventral aorta (G. *aorte* from *aeirein* = to lift): a large median vessel continuing forward from the conus of the heart. At the level of the anterior border of the metencephalon it will be observed in favorable specimens that the ventral aorta divides into two branches, the paired ventral aortae which continue to the level of the infundibulum where they turn dorsally and then posteriorly.

Dorsal aortae: The ventral aortae after looping dorsally and posteriorly become the dorsal aortae. The anterior parts of the dorsal aortae can not be seen owing to the thickness of the embryo. Posterior to the intestinal portal, however, they may be observed in good preparations of embryos with 14 or 15 somites as two very faint stripes, one on each side of the body lying beneath the somites;

their sides are marked by very thin lines. Note that caudal to the last somite they diverge laterally and give rise to a plexus of vessels on the blastoderm.

Vitelline (omphalomesenteric) arteries: the ends of the dorsal aortae just described and the adjacent plexus of vessels.

Vitelline circulation:

Vitelline arteries and veins: Trace on to the blastoderm the distal parts of the vitelline arteries and veins, the proximal segments of which were described above. Is there a connection between the arterial and venous channels?

Blood islands: Study the nature of the blood islands at this stage of development. What is their relation to the vitelline vessels?

Sinus terminalis (L. *sinus; terminus* = boundary): a wavy stripe encircling the outer margin of the area vasculosa. It marks a developing blood vessel which will become the terminal or bounding channel of the vitelline circulation.

Regions of the blastoderm:

Area pellucida, area (opaca) vasculosa, area (opaca) vitellina: as described earlier.

Proamnion: Note that the extent of the proamnion has now become considerably reduced. Trace, if possible, the anterior border of the mesoderm. What is its relation to the epimyocardium of the heart?

Subcephalic pocket: In very favorable preparations a thin line of ectoderm may be traced, as in the 23–25 hour embryo, under the embryo anterior to the conus region of the heart. This line, continuous with the ectoderm of the head and also that of the blastoderm, marks the posterior boundary of the subcephalic pocket.

109

2. Transverse sections of the 33 hour chick embryo:

Using the procedure followed in the exercise on the serial sections of the 24 hour embryo (see p. 94) identify and study with intermediate and high magnification the following structures.

a. Sections at the level of the prosencephalon and mesencephalon.

Structures essentially unchanged from condition in the 24 hour embryo:

Epidermis	Mesenchyme
Foregut	Area pellucida
Oral plate	Somatopleure
Proamnion	Coelom

Subcephalic pocket (trace to its posterior tip)

Prosencephalon: the forebrain; the large broad anteriormost division of the neural tube.

Prosocoel, optic vesicles, and *opticoels:* readily identified from description of whole mount.

Optic stalks: the tubular bridges connecting the optic vesicles to the prosencephalon proper.

Infundibulum: a shallow depression in the floor of the forebrain at the level of the posterior border of the optic vesicles.

Anterior neuropore: a median cleft at the anterior tip of the prosencephalon (seen only in the first one or two sections passing through the neural tube) or a notch in the dorsal or ventral wall of the neural tube. The cleft or notch means that the neural folds have not yet completely fused in this region.

Mesencephalon: the midbrain; an oval-shaped vesicle posterior to the prosencephalon. Its cavity is the *mesocoel.* Most of the mesencephalon is underlain with notochord, seen in cross section as a small irregular mass of cells.

Ventral aortae: Locate the section in which the ectoderm

of the head becomes fused to the ectoderm of the blasto-derm. Usually at this level one may observe two thin-walled tubes below the foregut, one on each side of a median de-pression in the floor of the foregut. These are the paired ventral aortae.

Dorsal aortae: in the same section a similar but larger pair of tubes lying dorsal to the foregut.

First aortic arches: Trace cranially the ventral and dorsal aortae just identified. Note that they become continuous by means of loops over the anterolateral aspect of the foregut. These loops are actually the first pair of aortic arches.

Stomodeum: a shallow midventral depression in the ecto-derm. The floor of the stomodeum together with the adja-cent entodermal evagination of the foregut constitutes the oral plate. The stomodeum is the forerunner of the buccal cavity.

Anterior cardinal veins (L. *cardo* = hinge or turning point, hence, important): a venous drainage of the head seen here as small spaces, lined with a thin endothelium, situated in the mesenchyme lateral to the neural tube.

Area vasculosa: Note that now the splanchnic mesoderm has formed blood vessels (vitelline vessels). Do the vessels contain blood corpuscles? Arteries and veins can not be differentiated.

b. Sections at the level of the heart.

Ventral aorta: Trace caudally the paired ventral aortae from the regions studied above until they fuse to give a median unpaired ventral aorta.

Conus of the heart: Continue to trace the ventral aorta caudally until the vessel seems to bulge ventrally into the coelom. This region of the system may now be regarded as the future conus. Note in detail:

Epimyocardium: the outer, heavier layer in the wall of the heart.

111

Endocardium: the inner, thinner layer in the wall of the heart.

Pericardial cavity (G. *peri* + *kardia*): the region of the coelom immediately surrounding the heart. Is it embryonic or extra-embryonic coelom? What is the relation of the two here? What is the origin of the pericardial cavity?

Dorsal mesocardium (G. *mesos* + *kardia*): the broad connection or isthmus between the heart and the foregut.

Ventricle: Trace the heart farther caudally. Note that it bends far to one side. This region of the tube may be regarded as the future ventricle. The dorsal mesocardium is narrower here. Is there a ventral mesocardium? What has become of it? Are you viewing the sections of the embryo as though you were at the anterior or posterior end of the embryo?

Rhombencephalon: At the level of the ventricle one observes the hindbrain, a thick-walled vesicle dorsal to the notochord. Is it metencephalon or myelencephalon? Review the whole mount on this point. Trace the neural tube anteriorly. Distinguish between myelencephalon and metencephalon; between metencephalon and mesencephalon.

Auditory placodes: Return to the sections showing the ventricle far to the right side of the midline. Note thickenings in the ectoderm, sometimes depressed, lying lateral to the hindbrain. These are the auditory placodes or pits.

Neural crests: small clusters of cells squeezed between the neural tube and the ectoderm (epidermis).

Thyroid: The first indication of the thyroid is to be seen in a median shallow depression of the foregut above the dorsal mesocardium.

Atrium: Proceed now to trace the heart posteriorly. When the tube begins to "swing back" toward the midline it may be stated that this is now atrium, the forerunner of the future auricles.

112

Sinus venosus: When the heart is again in the midline and is a transversely elongated, dorsoventrally flattened tube, the section is passing through the sinus venosus.

Anterior intestinal portal: In the next few posterior sections note that the foregut seems to break through the sinus venosus to open onto the yolk. This ventral opening of the foregut marks the anterior intestinal portal.

Vitelline (omphalomesenteric) veins: large tubes or vessels, one on each side of the anterior intestinal portal. Trace these back into the sinus venosus and (by proceeding caudally in the series) laterally onto the blastoderm.

Anterior cardinal veins: a pair of small vessels (clear oval spaces) above the dorsal aortae and adjacent to the ventrolateral aspect of the neural tube in sections from heart region of the embryo.

c. Sections through the somitic region.

Spinal cord: Observe its elliptically shaped cavity, the central canal, and under high power the finer details of its walls. Are neural crests present?

Notochord, somites, nephrotome, and *lateral plate:* as described for the 24 hour embryo (see p. 99).

Dorsal aortae: a pair of large vessels lying between the somites and the entoderm. Trace these vessels caudally to the level of the last somite. Note that the dorsal aortae have "migrated" laterally. Trace farther; observe that they extend laterad into a plexus of vitelline vessels. At this level they are designated *vitelline (omphalomesenteric) arteries.*

d. Sections through the sinus rhomboidalis.

Neural tube: Note the unfused, widely separated neural folds.

Notochord: Trace the notochord into Hensen's node.

Unsegmented mesoderm: Observe that the mesoderm is not yet cut up into somites and that laterally also subdivisions do not exist, i.e., somite, nephrotome, and lateral plate.

Primitive streak: Proceed to more posterior sections to observe the primitive streak. Note that immediately caudal to the level of Hensen's node the neural folds give way to the low primitive ridges. Has the picture of the primitive streak changed from that of the 24 hour embryo?

Vitelline vessels: Observe the vitelline vessels and their relationship to the blood islands.

Review Questions

1. Describe the formation of the chick heart from its primordia through the stage observed in the 33 hour embryo.

2. Why does the heart bend? What might explain its bending to the right? What is meant by differential growth? Name other examples.

3. What are the origin and fate of the blood islands? Where are erythrocytes, lymphocytes, and polymorphonuclear leucocytes produced in the adult? Are other embryonic organs hemopoietic (G. *haima* = blood + *poiesis* = making) i.e., blood forming?

4. Name the five major divisions of the adult vertebrate brain and their respective cavities.

5. Name some of the derivatives of the neural crests.

6. Describe the system of cardinal veins. How do these vessels, as examples of intraembryonic blood vessels, form in the chick?

7. Distinguish between an artery and a vein functionally and histologically.

8. In what respects would the blood in the vitelline veins differ from that in the vitelline arteries?

9. Diagram a cross section (A-B), a sagittal section (C-D), and a parasagittal section (E-F) of the embryo (fig. 37). Use blue pencil for ectoderm, yellow for entoderm, and red for mesoderm, with heavy lines for epimyocardium and a lighter line for endocardium. Label.

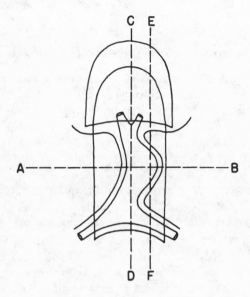

Fig. 37. Thirty-three hour chick embryo.

ORIGIN OF ASYMMETRY

At first vertebrate embryos are bilaterally symmetrical. They become asymmetrical owing to twisting of organs. But what causes the heart, for example, to bend to the right? Differential growth. The right sides of cardiac tubes from early chick embryos cultured in hanging drops grow more rapidly than the left sides, forcing the tubes to assume C shapes. In rare instances in vertebrate development the left side elongates more and the heart bends to the left. This anomaly is *situs inversus*. It usually involves a reversed asymmetry of the abdominal viscera as well. Spemann discovered that *situs inversus* of heart and gut could be produced by rotating a square of the archenteric roof in the neurula of a salamander. And in his experiment on constricting blastulae (but not 2-cell embryos) the lefthand twin was normal whereas the righthand twin often exhibited *situs inversus*. The inference is that there may be a center in the archenteric roof that determines the asymmetrical rates of growth. Sinistral or dextral coiling in certain snails depends on the tilt of the mitotic spindle in the early spiral cleavages, and this is determined by genes acting in the *oocyte* (an example of maternal inheritance).

D. Development in the 48 Hour Chick Embryo

1. WHOLE MOUNT OF THE 48 HOUR CHICK EMBRYO: Examine under low and intermediate power only. In studying certain structures such as the heart a ventral view is helpful. Carefully support the inverted slide on a couple of coins placed on the stage of the microscope. Identify the following features.

Flexion and *torsion:* bending and twisting of the embryo.

Cranial flexure: a sharp ventral bending of the head of the embryo through almost 180 degrees. The anterior end of the embryo thus becomes an inverted U.

Cervical flexure: a gentle curvature of the embryo in the future neck or cervical region. Cranial and cervical flexures are segments of a general ventral bending of the embryo. Later more caudal segments will be involved and the entire embryo will be curved to form a large C.

Torsion: a rotation of the body about its long axis, beginning at the anterior end, so that the embryo comes to lie on its left side. In the 48 hour embryo how far caudally has torsion progressed?

PATTERNS OF GROWTH

You have probably wondered why the chick embryo undergoes the flexures and torsion just observed. Can you think of any advantage to the chick embryo of cranial and cervical flexure? of lying on its left side? Because causal factors may be less obvious, the author suggests the following.

1. *Spatial limitations.* Physical restrictions are imposed by investments (e.g., leathery or calcareous shells of fishes, reptiles, birds, and monotreme mammals), by incompressible liquids (e.g., yolk, albumen, blood, and cerebro-spinal fluid), and by large organs (e.g., liver, brain, and heart). In the instance of the cerebro-spinal fluid, which is secreted into the neurocoel, the brain and spinal cord are under internal hydrostatic pressure which is surely a morphogenic factor. Blood pressure is a tremendously important molding force in the development of the vascular system, and many examples of this so-called hemodynamic principle will be encountered later.

116

2. *Differential rates of growth.* Certain organs have higher rates of proliferation at specific times of development than other organs (e.g., neural tube early in development or liver in later stages) or given regions of an organ or system may have a higher mitotic index than adjacent regions (e.g., medulla versus spinal cord). The extension of the notochord is probably of importance in the axial elongation of the neural tube. The configuration of the tube, on the other hand, appears to be due to the differential developmental properties of each level (anterior to posterior and dorsal to ventral). The tube is a mosaic of small fields, each with its intrinsic properties, one of which is rate of growth. The consequences of localized growth pressures are invaginations, evaginations, bendings, thickenings, elongations, and shifts of relationships of districts to one another.

3. *Cell loss.* Death, phagocytosis, emigration, etc. of cells are also natural and important factors in giving form to an organ or its parts.

Brain:

Mesencephalon: The midbrain is now the part of the embryo seen first owing to cranial flexure and the consequent bending of the forebrain ventrally and posteriorly. Its boundaries are marked by the following.

Isthmus (G. *isthmos* = narrow passage) or *mesometencephalic fold:* a prominent circumferential constriction in the brain marking the boundary between mesencephalon and hindbrain.

Meso-diencephalic fold: a shallow constriction in the brain, largely on the dorsal surface, establishing a boundary between mesencephalon and forebrain.

Tuberculum posterius (L. *tuberculum* = a little hump or swelling): a thickened elevation in the floor of the brain, about opposite the meso-diencephalic fold, the anterior margin of which marks the ventral boundary between midbrain and forebrain.

Prosencephalon: now divided into two subregions.

Diencephalon (G. *dia* = through or between + *enkephalos*): the posterior subdivision of the forebrain,

117

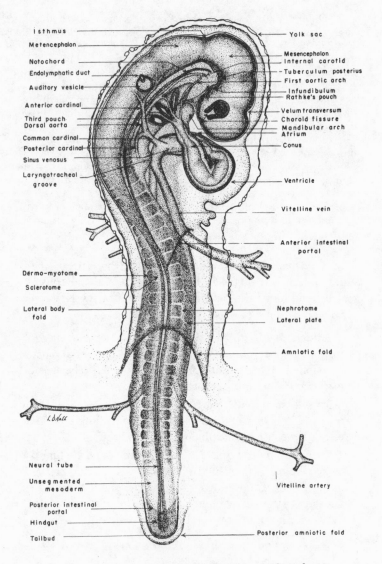

Isthmus — Yolk sac

Metencephalon —

Notochord —
 Mesencephalon
 Internal carotid
Endolymphatic duct — Tuberculum posterius
 First aortic arch
Auditory vesicle — Infundibulum
 Rathke's pouch
Anterior cardinal — Velum transversum
Third pouch — Choroid fissure
Dorsal aorta — Mandibular arch
Common cardinal — Atrium
Posterior cardinal — Conus
Sinus venosus —

Laryngotracheal
 groove — Ventricle

 Vitelline vein

 Anterior intestinal
 portal

Dermo-myotome —

Sclerotome —

Lateral body —
 fold
 Nephrotome
 Lateral plate

 Amniotic fold

Neural tube —

Unsegmented —
 mesoderm
 Vitelline artery
Posterior intestinal —
 portal
Hindgut —
Tailbud — Posterior amniotic fold

Fig. 38. Whole mount of 48-50 hour chick embryo.

118

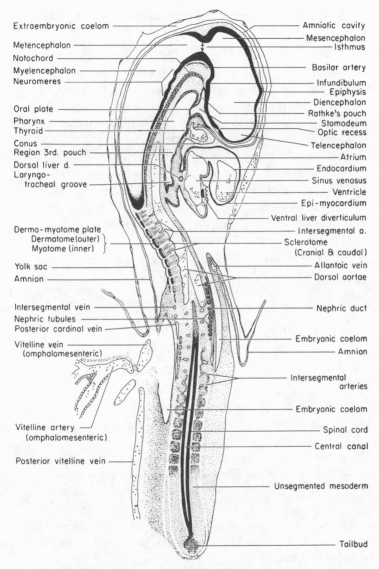

Fig. 39. Sagittal section of 48 hour chick embryo.

119

the posterior boundary of which has just been stated as the meso-diencephalic fold and the tuberculum posterius. Its anterior limits are marked by:

Velum transversum (L. *velum* = awning; *transvertere* = to direct across): a depression in the dorsal wall of the forebrain about midway between the meso-diencephalic fold and the anterior end of the brain; the demarcation between diencephalon and telencephalon.

Optic recess: a slight notch in the ventral lining of the forebrain about opposite the velum transversum, marking the future ventral boundary between diencephalon and telencephalon. Careful focusing is required to discern this feature.

Infundibulum: shallow depression in the floor of the diencephalon, a short distance anterior to the tuberculum posterius. Better seen in embryos of 50 hours incubation.

Telencephalon (G. *tele* = far + *enkephalos*): the region of the forebrain anterior to the velum transversum and the optic recess.

Metencephalon: a narrow poorly defined region of the hindbrain immediately following the mesencephalon. Anteriorly it is delimited by the isthmus. The point at which the dorsal wall of the hindbrain becomes very thin may be taken as the dorsal boundary between metencephalon and myelencephalon. The ventral boundary is not marked.

Myelencephalon: the large brain vesicle posterior to the metencephalon, characterized by the thin roof which later receives many blood vessels and sinks into the myelocoel to form the *posterior choroid plexus*. Can neuromeres be seen? How many?

Eyes: The optic vesicles seen in the 33 hour chick as

simple evaginations of the diencephalic region of the forebrain now show additional features.

Optic cups: elliptical or pear-shaped structures on each side of the forebrain, derived from the optic vesicles by invagination (best seen in serial sections). By careful focusing identify right and left optic cups. In older specimens the cup can be seen to be double layered (further details to be noted in the serial sections).

Lens vesicle: Within each optic cup may be seen a faint thick-walled circle, the developing lens.

Optic stalk: From the ventral wall of each optic vesicle a tube, the optic stalk, extends medially toward the brain. It appears in optical section (a plane passing through it) as a small thick circle. The optic stalks connect to the brain at the boundary between diencephalon and telencephalon.

Inner ear:

Auditory vesicles or *otocysts* (G. *otos* = ear + *kystis* = sac): ovoid vesicles, one on each side of the myelencephalon, derived by invagination from the ectoderm.

Endolymphatic duct (G. *endon* + L. *lympha* = water): In older specimens the dorsal wall of the vesicle has a small evagination, the forerunner of the endolymphatic duct.

Digestive tract:

Pharynx: the region of the digestive tube roughly dorsal to the heart and ventral to the myelencephalon. The pharynx is quite flat dorsoventrally but wide laterally owing to paired outpocketings known as pharyngeal pouches. Corresponding to the pouches of entoderm there are ectodermal invaginations, termed pharyngeal grooves or furrows. The plate of fused entoderm and ectoderm is called the closing plate. If the closing plate is broken the perforation is a pharyngeal cleft.

Second pharyngeal pouch: prominent dilatation in the pharynx situated at the level of the posterior border of the auditory vesicle. The lateral wall of the pouch appears as a vertical light streak owing to the existence of an actual pharyngeal cleft in older specimens or an extremely thin closing plate.

First or *hyomandibular (pharyngeal) pouch:* usually less prominent than pouch 2, appearing as an oblique light streak anterior to the second pouch. The light streak, representing the closing plate, is quite long, extending from the anterior border of the thyroid anterodorsally to almost the level of the neural tube.

Third pharyngeal pouch: usually a faint streak or ovoid area, indicating the closing plate of the third pouch, vertically oriented and situated posterior to the second pouch.

Visceral arches: The mesenchyme between any two pouches constitutes a visceral arch. The first arch, anterior to the first pouch or cleft, is a large mass rounded ventrally and is designated the *mandibular* arch since it forms the jaws. The second arch, between the first and second pouches, is the *hyoid* (G. from the letter upsilon + *eidos* = from). The third arch lies between the second and third pouches, the fourth arch between pouches 3 and 4, and the fifth arch, yet to be differentiated, behind the fourth pouch. Arches posterior to the hyoid are sometimes designated branchial arches because in the lower vertebrates they bear gills (see frog larva, fig. 17).

Thyroid: a median depression in the floor of the pharynx at the level of the anterior border of the second pouches.

*Laryngotracheal groove:** In specimens of 50 hours incubation or more the foregut behind the pharyngeal pouches is higher dorsoventrally. This is due to a groove or ventral outpocketing, the laryngotracheal groove, which later develops larynx, trachea, and lung buds.

Esophagus: ° The future esophagus will later differentiate from the part of the foregut dorsal to the laryngotracheal groove.

Anterior intestinal portal: identified by a heavy arched line at the level of the caudal border of the heart. Owing to torsion the arched line is no longer transversely, but now obliquely, oriented. The foregut between the intestinal portal and esophagus will become stomach and duodenum.

Stomodeum: an ectodermal invagination anterior to the mandibular arch, appearing in lateral aspect as a narrow light streak or cleft between the head and the mandibular arch, marking the position of the future mouth.

Preoral gut: the anteriormost tip of the foregut which actually extends forward as a fingerlike diverticulum anterior to the future mouth.

Rathke's pouch: By careful focusing a diverticulum of the stomodeum may be identified extending underneath the head toward the infundibulum. This pocket is short in younger specimens, longer in older ones. Rathke's pouch plus the infundibular floor will become the pituitary body.

Notochord: faint medial line extending forward beneath the central nervous system almost to the level of the tuberculum posterius. It is best seen in the space between preoral gut and hindbrain. The anterior tip of the notochord beneath the midbrain is bent gently by the cranial flexure of the embryo.

Somites: How many have differentiated? Note that some of the trunk somites show subdivisions.

Dermo-myotome: outer dense plate of cells which becomes the *dermatome* (G. *derma* = skin + *tomos* = a cutting, hence, a section) and the *myotome* (G. *myos* =

muscle + *tomos*), forming respectively dermis and muscle.

Sclerotome (G. *skleros* = hard + *tomos*): inner loosely arranged cells, the forerunner of vertebrae.

Heart: large double-walled organ lying in the concavity of the embryo created by cranial and cervical flexures. The heart, coiled as shown in figure 38, may be analyzed into the following regions.

Conus: at high focus a straight somewhat narrow chamber of the heart lying parallel with or obliquely to the hindbrain. Identify the outer *epimyocardium* and the inner *endocardium*.

Ventricle: at high focus a large vesicle continuous with the conus extending ventrally at right angles to the embryo. Frequently the chamber inside the endocardium will appear granular. What are these "granules"?

Atrium: at low focus a curved chamber of the heart lying anterior to the ventricle, with which it is continuous, and arching under (to the left of) the conus. Trace the continuity of the epimyocardium and endocardium of ventricle and atrium.

Sinus venosus: at low focus a small chamber posterior to but continuous with the atrium and lying beneath (to the left of) the junction of ventricle and conus.

Blood vessels: Blood channels can be identified in the whole mount by masses of corpuscles which fill them or in areas where the preparation is dark (thick) as faint light streaks. After examining the blood vessels on the whole mount, review the system in the demonstration of chick embryos injected with india ink.

Ventral aorta: can not be seen in younger specimens and only with some imagination in older ones, as a poorly defined short vessel running from the conus to the floor of the pharynx. It gives rise immediately to the aortic arches.

Aortic arches: best seen in older specimens, those of 50 hours incubation or more. By careful focusing observe that the aortic arches are paired, one member of each pair on the right and left sides of the pharynx.

First arch: faint light streak passing dorsally, somewhat obliquely, through the first pharyngeal arch, then anteriorly around the preoral gut as the *carotid loop.*

Second arch: usually clear stripe, running vertically toward the auditory vesicle through the second pharyngeal arch.

Third arch: narrow very faint streak passing dorsally through the third pharyngeal arch.

Dorsal aortae: large vessels, right and left, running posteriorly along the dorsolateral wall of the pharynx. Note their relation to the aortic arches, especially the first pair. Posterior to the pharynx and in the region of the cervical flexure the dorsal aortae, in older specimens, are fused to form a large median unpaired dorsal aorta. Further tracing posteriorly is difficult.

Vitelline (omphalomesenteric) arteries: Approximately at the level of the 22d somites the dorsal aortae—now paired again—give rise on each side to large vitelline arteries which run out laterally to the vitelline system on the blastoderm.

Vitelline circulatory system: Examine the vitelline vessels on the blastoderm. Observe the walls of the arterial system in the area vasculosa. Trace certain branches into the venous drainage. Note *sinus terminalis* described earlier.

Vitelline (omphalomesenteric) veins: Follow the system of veins to the right of the embryo converging to form the right vitelline vein. Trace it mesiad to the sinus venosus of the heart. The proximal segment of the left vitelline vessel is obscured by the embryo; note its distal distribution. Observe that anterior and posterior vitelline veins are forming in older embryos. These drain into the left vitelline vein which becomes much larger than the right. Low-power magnification is useful in this study.

Anterior cardinal vein: With carefully adjusted focus find a light broad streak arching down from the post-otic region toward the laryngotracheal region of the gut. This is the anterior cardinal vein. Locate both right and left vessels. Trace them as far anteriorly as possible.

Posterior cardinal vein: At the level of the laryngotracheal groove note that a somewhat smaller very faint vessel coming from the trunk of the embryo enters the anterior cardinal vein. This is the posterior cardinal vein.

Common cardinal vein° or *duct of Cuvier* (after Georges Cuvier, French comparative anatomist, 1769–1832): the united anterior and posterior cardinal veins; a short vessel which continues ventrally to empty into the sinus venosus of the heart. It is seen best in injected embryos.

Dorsal intersegmental vessels:° smaller arteries and veins intersegmentally arranged. The arteries run from the dorsal aorta to dorsal organs such as neural tube, whereas the veins drain these organs into the cardinal veins.

Tail bud: solidly staining mass at the posterior tip of the embryo, marking the position of the old primitive streak and Hensen's node.

Tail fold: In older specimens (50 hours incubation) the tail bud is beginning to grow caudally over the blastoderm creating a tail fold which although formed somewhat differently is like the head fold. Observe that the ectoderm of the tail is seen as a U-shaped line.

Hindgut: a sacculation of the entoderm into the tail bud appearing as a U-shaped area, the margins of which are parallel to the ectoderm of the tail.

126

Posterior intestinal portal: the broad arched opening of the hindgut into the yolk.

Amniotic folds:

Anterior fold: Usually several segments behind the anterior intestinal portal may be seen the anterior amniotic fold arching across the dorsum of the embryo. This fold represents the posterior margin of a hood growing caudally over the embryo. It is composed of two membranes: an inner one, the *amnion;* and an outer one, the *chorion.* Each membrane in turn is composed of two layers: ectoderm and somatic mesoderm (to be seen later in the serial sections).

*Posterior fold:** In older specimens a suggestion may be seen of a posterior hood forming from the blastoderm and beginning to enclose the tail of the embryo, namely, a faint streak or fold posterior to the tail.

2. DEMONSTRATION OF CIRCULATION IN THE 48 HOUR CHICK EMBRYO: The shell of an egg incubated for 48 hours is gently cracked and then removed under warm normal salt solution. The blastoderm is carefully cut with scissors outside the area vasculosa so that the embryo may be floated free of the yolk and into a watch glass for observation under the microscope.

Study the contractions of the various chambers of the heart and the circulation of blood through the embryonic and extraembryonic systems.

Examine also chick embryos in which the circulatory system has been injected with india ink. These preparations will permit the observation of many vessels not easily seen in the fixed material.

3. TRANSVERSE SECTIONS OF THE 48 HOUR CHICK EMBRYO: Owing to flexion and torsion the first sections through the embryo will be frontal ones. Reference to the whole mount should be made frequently and the use of a straightedge on figure 38 will be helpful in understanding the plane of section-

127

ing. Handle the preparations with care, being particularly cautious in lowering the stage clips on the slides. Sudden pressure will seriously damage the sections. Identify the following structures.

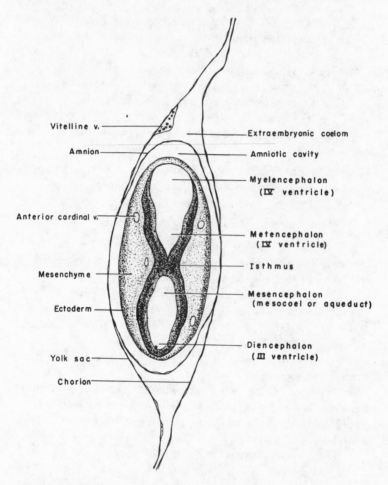

Fig. 40. Cross section through extreme anterior end of 48 hour chick embryo.

Fetal membranes: Select any section passing through the anteriormost part of the embryo.

Amnion: double-layered membrane completely encircling the section of the embryo, and consisting of an inner layer of ectoderm and an outer layer of somatic mesoderm. The amniotic cavity lies between the amnion and the embryo.

Yolk-sac: a double-layered membrane, identified by the presence of blood vessels, consisting of splanchnic mesoderm (nearest to the embryo and containing the blood vessels), and a layer of entoderm. To which side of the embryo does the yolk-sac lie?

Chorion: a double-layered membrane of an inner layer of somatic mesoderm adjacent to the amnion and an outer layer of ectoderm. It lies to the right of the embryo.

Proamnion: a persisting small region of the blastoderm beneath (to the left of) the embryo into which mesoderm has not yet extended.

Extraembryonic coelom: the space, bounded by mesoderm, between the chorion, yolk-sac, and amnion.

Mesencephalon: the thick-walled vesicle appearing in the first sections passing through the brain of the embryo. Verify this by reference to the whole mount. What is the cavity?

Mesenchyme: the loose reticulum between the epidermis and the brain.

Brain in frontal section: Trace farther caudally in the series until the picture of the brain is that of a "dumbbell." Here the section is passing frontally through several divisions of the brain as follows.

Isthmus: the constriction at the middle of the "dumbbell."

Myelencephalon: the end of the "dumbbell" with the thin roof.

Metencephalon: a narrow undefined region between the isthmus and the myelencephalon.

Mesencephalon: region of the brain on the other side of the isthmus, corresponding in position to the metencephalon.

Diencephalon: the end of the "dumbbell" opposite the myelencephalon; possesses a moderately thick roof with a small notch in the midline, representing the beginning of an evagination to form the *epiphysis* or *pineal gland.*

All of the above features may or may not be shown in a given section, depending upon the plane of sectioning. Trace the "dumbbell" caudally in the series until the isthmus loses its cavity. Refer to the whole mount to understand this picture. Trace farther until the figure appears to "break" into two sections of brain: one with a thin roof, the myelencephalon; the other, the diencephalon. What are the two cavities?

Notochord: In the sections immediately following those just studied note that a small elongated dark mass appears between the myelencephalon and diencephalon. This is the notochord. Tracing farther this mass separates into two pieces. Explain. What becomes of these pieces in more posterior sections?

Anterior cardinal veins: Return now to the section showing the tuberculum posterius. Note a space in the mesenchyme along each side of the hindbrain. These are the anterior cardinal veins. Trace them caudally in the series. They become very long empty spaces extending forward toward the diencephalon. In the next few sections they break up into a number of vessels. Explain this picture.

Fifth cranial nerve or *trigeminal:* In the sections just studied showing the anterior cardinal veins note that from the ventrolateral wall of the hindbrain there emerges on each side a dark mass which extends toward the skin immediately lateral to the cardinal vein. This is the ganglion of the fifth nerve.

Carotid loops: Medial to the anterior cardinal veins will appear two vessels, seen as elongated spaces usually containing blood corpuscles. These are the carotid loops, extensions of the first aortic arches, which loop around the anterior end of the gut and connect with the dorsal aortae.

Preoral gut: [*] a small circle or oval seen between the carotid loops. This is the preoral gut or the anteriormost part of the foregut.

Infundibulum: an evagination from the floor of the forebrain extending in the direction of the foregut.

Telencephalon: The dorsal part of the section of the brain containing the infundibulum is probably now telencephalon. Verify this point on the whole mount.

Pharynx: The preoral gut has now expanded into the pharynx which exhibits a three-armed figure, the medial arm being the pharynx proper and the right and left arms the *first pharyngeal pouches.* How prominent are the pharyngeal furrows? Have the closing plates ruptured yet?

Rathke's pouch (after Martin Heinrich Rathke, German anatomist, 1793–1860): an ectodermal tubule between the infundibulum and the foregut.

Internal carotid arteries: paired vessels extending from the carotid loops alongside the forebrain and medial to the optic cups, structures to be described later.

Dorsal aortae: The first pharyngeal pouches seemed to "break" the carotid loops into two segments. The vessels lying dorsal to the pouches on either side of the notochord are hereafter the dorsal aortae. The vessels ventral to the pouches are the first aortic arches. Verify this point on the whole mount.

Acoustico-facialis ganglion of the *VII* and *VIII cranial nerves:* small dark bodies lateral to the myelencephalon and anterior cardinal veins.

Stomodeum: Deep clefts now seem to form on each side, in more caudal sections, and to fuse in the midline. This is

131

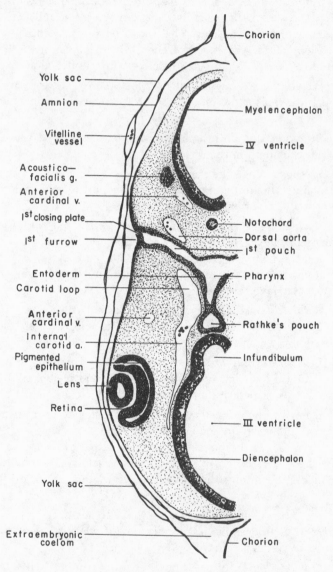

Fig. 41. Cross section of 48 hour chick embryo at level of Rathke's pouch.

132

the stomodeum. Note that Rathke's pouch opens into it.

Oral plate or *pharyngeal membrane:* the thin bar (ectoderm and entoderm) separating the stomodeum and the pharynx.

First visceral or *mandibular arch:* the large masses of mesenchyme, one to each side of the oral plate. Each mass or half of the arch contains a vessel, a member of the *first pair of aortic arches.*

Auditory vesicles or *otocysts:* paired vesicles alongside the myelencephalon; note that these vesicles arise as invaginations of the ectoderm.

Optic cup: double walled cup to each side of the forebrain.

> *Presumptive retina:* inner layer of the cup. Compare its histology with that of the brain.

> *Presumptive pigmented epithelium:* outer layer of the cup which will later become the pigmented layer of the retina.

Lens: vesicle within the concavity of the optic cup. Note that the lens vesicle arises as an invagination of the ectoderm.

Optic stalks: the connections between the optic cups and the forebrain (diencephalon still in the ventral part of the section; telencephalon above).

Ventral aorta: The bases of the first aortic arches, sometimes called ventral aortic roots, now "fuse" to form the ventral aorta beneath the pharynx.

Second aortic arches: in more posterior sections two large vessels passing ventrally from the dorsal aortae. Trace them into the ventral aorta (counter to flow of blood in them, of course). They pass through the substance of the second or hyoid visceral arch.

Ganglion of the *IX cranial nerve* or *glossopharyngeal:* paired dark bodies immediately posterior to the auditory vesicles.

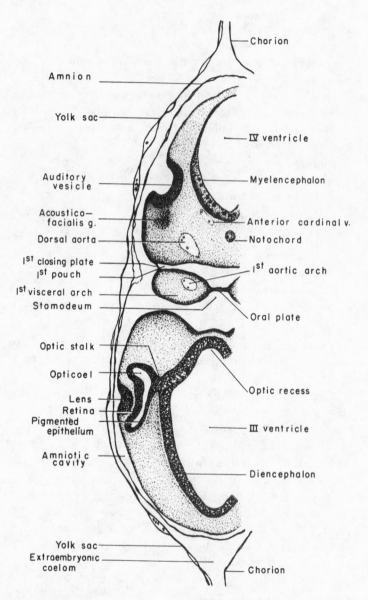

Fig. 42. Cross section of 48 hour chick embryo at level of oral plate.

Second pharyngeal pouches: lateral sacculations of the pharynx which seem to break through the second aortic arches. Note that these pouches extend directly laterally instead of dorsolaterally as do the first pouches.

Third aortic arches: a pair of blood vessels, smaller than the second arches, extending around the pharynx from the dorsal aortae to the ventral aorta. They pass through the substance of the third visceral arch.

Thyroid: V-shaped diverticulum from the floor of the pharynx.

Ganglion of the *X cranial nerve* or *vagus:* faint masses of cells, one on each side, lateral to the myelencephalon and above the anterior cardinal veins.

Nasal placodes: slight thickenings in the ectoderm lateral to the telencephalon.

Third pharyngeal pouches: large sacculations of the pharynx immediately following the third aortic arches. Pharynx and pouches here give the foregut the aspect of a large oval transversely oriented.

Conus: With the entrance (actually exit) of the third aortic arches into the ventral aorta the latter is seen to be continuous with the anteriormost chamber of the heart, the conus, which bulges ventrally into the coelom. Here the future pericardial cavity of the embryonic coelom is broadly continuous with the extraembryonic coelom. Identify endocardium and epimyocardium.

First somite: At the level of the anterior end of the conus or even farther forward may be seen the first somite.

Dorsal mesocardium: the mesodermal bridge connecting the heart (conus) to the dorsal wall of the coelom; to be seen again in the sinoatrial region.

Atrium: Continuing caudally in the series, another chamber of the heart, the atrium, appears to the left of the conus.

Dorsal aorta: Note that the paired dorsal aortae have fused.

135

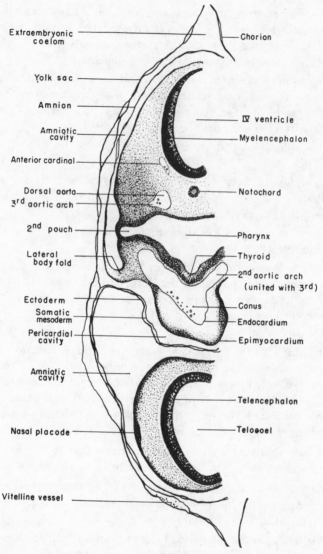

Fig. 43. Cross section of 48 hour chick embryo at level of second pouch.

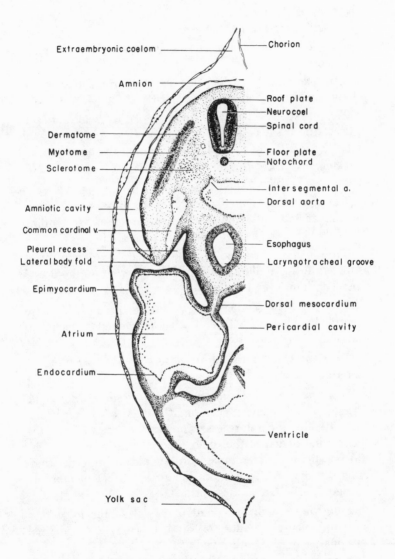

Extraembryonic coelom ——————— Chorion

Amnion ——————

Roof plate
Neurocoel
Spinal cord

Dermatome ——————
Myotome ——————
Sclerotome ——————

Floor plate
Notochord

Intersegmental a.
Dorsal aorta

Amniotic cavity ——————

Common cardinal v.——————
Pleural recess ——————
Lateral body fold ——————

Esophagus

Laryngotracheal groove

Epimyocardium——————

Dorsal mesocardium

Pericardial cavity

Atrium ——————

Endocardium——————

Ventricle

Yolk sac ——————

Fig. 44. Cross section of 48 hour chick embryo at level of heart.

137

Laryngotracheal groove: a deep V-shaped depression in the floor of the foregut. The dorsal part of the gut may be regarded as the future esophagus.

Anterior cardinal veins: Note that these vessels are "moving" ventrad.

Ventricle: the large loop connecting atrium and conus; the ventralmost chamber of the heart.

Pleural recesses: fingerlike recesses of the coelom to each side of the tracheoesophageal region of the gut. They will later become the pleural cavities enclosing the lungs.

Common cardinals: elongated vessels, one on each side, lateral to the pleural recesses.

Dorsal intersegmented arteries: Note these small vessels arising at intervals from the dorsal aorta and extending dorsally between the spinal cord and the somite.

Sinus venosus: the posteriormost chamber of the heart situated in the midline and broadly supported by the dorsal mesocardium.

Ventricle: The section is now passing through the large ventricular chamber which lies below and unconnected to the sinus.

Ducts of Cuvier: the large vessels (*common cardinals*) entering into the sinus venosus.

Pleuropericardial membranes (lateral mesocardia): the bridges of mesenchyme enclosing the ducts of Cuvier which separate the pleural recesses (pleural cavities) from the pericardial region of the coelom. Some authors make a distinction between lateral mesocardia (folds bearing the common cardinal veins) and the pleuropericardial membranes (ridges on the lateral mesocardia).

Lung buds: slight evaginations of splanchnic mesoderm into the pleural cavities. A foreshadowing of entodermal evaginations may be observed also.

Posterior cardinal veins: Passing farther caudally note that

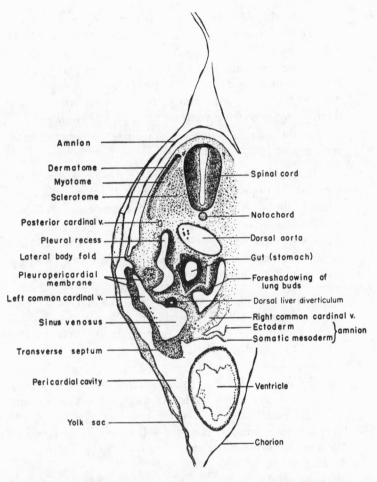

Fig. 45. Cross section of 48 hour chick embryo at level of common cardinals.

from the dorsal tip of the common cardinal there persists a vessel appearing as a circular space; this is the posterior cardinal vein.

Transverse septum: the mesenchyme surrounding the

139

sinus venosus which forms an incomplete transverse partition. To it are fused the pleuropericardial membranes identified above.

Vitelline (omphalomesenteric) veins: paired vessels which appear in place of the sinus venosus as one proceeds caudally in the series. Trace them; note that first the left and then the right vitelline veins pass out onto the yolk sac.

Liver diverticula: Return to the sections through the transverse septum.

Dorsal diverticulum: In the chick there is a dorsal outpocketing appearing as an oval or circle lying in the transverse septum and dorsal to the sinus venosus.

Ventral diverticulum: a bilobed outpocketing (sometimes its bilobed character is not clear) situated below the vitelline veins. Trace the connections of the two liver diverticula to the gut.

Anterior intestinal portal: After the liver diverticula have united with (actually originated from) the gut, note that the laterally compressed foregut opens onto the yolk via the anterior intestinal portal.

Lateral body folds: All along the series it may be seen that the embryo is being undercut from the sides by two folds, the lateral body folds. At the level of the anterior intestinal portal they bear vessels which can be traced into the posterior cardinals. The groove between the embryo and the amnion formed by these folds is designated the *lateral limiting sulcus*.

Torsion: Observe that posterior to the anterior intestinal portal the embryo "rights itself" into a dorsoventral relation to the yolk-sac instead of lying with its left side upon it. This means, of course, that the sections are passing posterior to the region of torsion.

Midgut: the floorless gut between the anterior and posterior intestinal portals.

Posterior margin of the *anterior amniotic fold:* Continue in the series until the amnion and chorion above the embryo seem to break. This marks the posterior boundary in the mid-dorsal line of the anterior amniotic fold. Just before the "break" the membranes are fused; this point is called the *chorioamniotic raphe.* Refer to the whole mount. Review the fetal membranes.

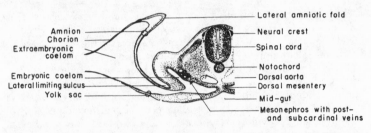

Fig. 46. Cross section of.48 hour chick embryo through mid-trunk region.

Lateral amniotic folds: As one continues farther caudally in the series note that on each side of the embryo folds of somatopleure form an inner amnion and an outer chorion. These lateral folds are merely lateral continuations of the anterior amniotic fold.

Dorsal aortae: The dorsal aorta has become paired again.

Mesonephros: Select a favorable section for study under high power.

Mesonephric ducts: small tubes lateral to the dorsal aorta and ventral to the posterior cardinal veins. These ducts were formed by delamination from the nephrogenic cord (nephrotome).

Mesonephric tubule: strand of cells extending mesiad from the mesonephric duct. Under high power and with a favorable preparation this strand will be seen to be composed of a double layer of cells with a tiny cleft between,

141

thus making a tubule. Note its entrance into the meso-nephric duct.

Nephrostome: Tubules may be found with openings into the coelom; these openings are nephrostomes. What is the function of a nephrostome?

THE VERTEBRATE NEPHRIC SYSTEM

The excretory organs of vertebrates have long been regarded as distinct members of a tripartite system of pronephros, mesonephros, and metanephros. In general the pronephros is the excretory apparatus of larval anamniotes, the mesonephros of adult anamniotes and embryonic amniotes, and the metanephros of adult reptiles, birds, and mammals. The pronephros is present in only rudimentary form in amniotes and in those anamniotes without larvae such as the sharks. The advocates of the concept of the developmental recapitulation of evolutionary history have stressed the parallelism between the above sequence in the phylogeny of the vertebrates and the gradients in time, position in the body, and structural complexity observed in the development of the nephric system in a higher vertebrate, such as a mammal. The first nephric units to form, albeit rudimentary and functionless, arise in the anteriormost (pronephric) segments of the nephrotome or mesomere; later the mesonephric region differentiates; and lastly the definitive organ, the kidney or metanephros, is formed. The spatial gradient in the chick, for example, has been given as follows: pronephros: somite levels 5-15; mesonephros: somite levels 13-30 (some overlap with the pronephros is said to occur); metanephros: somite levels 31-33. The structural gradient is one of increased length and complexity of the nephrons from the simple pronephric tubules—basically one per segment opening into the coelom (see fig. 18)—to the highly complex metanephric unit consisting of long, highly coiled tubules —many per segment, each with an internal glomerulus, and each differentiated into segments (proximal convoluted tubule, descending and ascending limbs of Henle's loop, distal convoluted tubule) and performing specific roles in excretion and selective reabsorption. The mesonephros occupies a somewhat intermediate position with respect to complexity of organization.

In contrast to the idea of a tripartite system is the concept of the holonephros (G. *holos* = whole + *nephros*) which holds that the vertebrate nephros is a single entity with gradations as noted above but without distinct boundaries between pronephric, mesonephric, and metanephric regions. Theodore Torrey has recently summarized the evidence in favor of the second interpretation; some of his points are as follows.

1. In the hagfish *Bdellostoma* nephric tubules develop at somite levels 11 to 82. Although at opposite ends of the series the nephrons are different (anterior ones having nephrostomes but no glomeruli; posterior ones possessing glomeruli but lacking nephrostomes), the intermediate zone reveals an insensible gradation between the two types.

2. The caudal end of the nephros of the shark, which under the tripartite system is termed a mesonephros, actually exhibits a high degree of complexity and in some instances has its own separate duct or ureter. A similar situation is found in certain bony fishes and amphibians. Torrey states that in instances of "simultaneous occurrence and/or structural continuity of the regions equivalent to meso- and metanephros, the organ is better described by the term opisthonephros" (G. *opisthen* = behind + *nephros*).

3. In amniote development, the chick embryo for example, the transition from the last "mesonephric" tubules to the first prospective "metanephric" tubules is imperceptible.

4. The basic morphogenic features are similar in all nephrons: small masses of intermediate mesoderm hollow to form a vesicle; the vesicle elongates into a tube; one end of the tube becomes connected to the nephric duct or a bud from it; the other end of the tube either opens into the coelom or forms a capsule which encloses a glomerulus.

5. Experimental embryology has shown that an exchange of parts of the mesomere (as between mesonephric and pronephric regions) leads to regulatory or neighborwise development. There are, therefore, no differences in morphogenic potency along the antero-posterior axis of the mesomere.

6. There is a fundamental uniformity in the physiology of all nephrons.

Somite: Select a section passing through a somite; examine under high power.

Dermo-myotome plate: thick plate consisting of two layers: an outer layer of columnar cells, the dermatome; and an inner layer, the myotome, which is quite thick dorsally but thins out ventrally.

Sclerotome: a body of loose cells medial to the dermo-

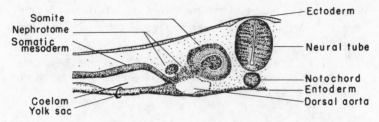

Somite
Nephrotome
Somatic mesoderm
Ectoderm
Neural tube
Notochord
Entoderm
Dorsal aorta
Coelom
Yolk sac

Fig. 47. Cross section of 48 hour chick embryo through posterior trunk region.

myotome plate and adjacent to spinal cord and dorsal aortae. Compare histologically these cells (scleroblasts) with the dermoblasts, the myoblasts, and the mesenchyme. At the same time look at the finer details of the neural tube, notochord, epidermis, etc. Observe that all blood corpuscles are nucleated.

Myocoel: the cavity of the somite; varies in size.

Dorsal mesentery: the median mesodermal bridge between the roof of the midgut and the dorsal organs.

Neural crests: small patches of cells at the dorsolateral margins of the spinal cord.

Vitelline (omphalomesenteric) arteries: No significant change takes place in the picture until the dorsal aortae are seen to extend onto the yolk-sac as the vitelline arteries. Do the dorsal aortae continue caudal to this point? At what level of the body do the vitelline arteries arise?

Nephrotome or *mesomere* or *intermediate mass:* Examine a section in which the nephrotome is not organized into mesonephric tubules. It consists of a loose aggregation of cells lateral to the somite and ventral to the mesonephric duct. Is the somite differentiated at this level? Examine the roof of the midgut. How far posteriorly can the mesonephric ducts be traced?

Unsegmented mesoderm: Examine a section in which the mesoderm is not segmented into somites.

Tail bud: Trace the neural tube, notochord, and unseg-

144

mented mesoderm into the tail bud. In what respect does the tail bud resemble the former primitive streak? Note irregularities in the terminus of the central canal.

Hindgut: At the posterior tip of the tail observe that the gut acquires a floor. This part of the gut is the hindgut. Is it long?

Posterior intestinal portal: the opening of the hindgut onto the yolk.

Anal plate: a fusion of the ectoderm and entoderm of the hindgut in the mid-dorsal line. Later this anal plate will be folded ventrally by the growth of the tail bud and the establishment of a tail fold. Like the oral plate it will rupture and form thereby the anal opening.

Review Questions

1. Name the landmarks on the dorsal surface of the neural tube used as boundaries between the divisions of the brain.

2. How many cranial nerves are present in amniotes? in the anamnia? Name the cranial nerves; which are sensory? motor? mixed?

3. Define the following terms pertaining to the pharynx: visceral arch, branchial arch, hyoid arch, pharyngeal groove or furrow, pharyngeal pouch, closing plate, oral plate, aortic arch.

4. What is the fate of Rathke's pouch? stomodeum? preoral gut?

5. What is the homologue of the first pharyngeal pouch in the shark? in a mammal?

6. Name some derivatives of the visceral arches and of the pharyngeal pouches.

7. How many aortic arches are represented in the ancestral vertebrate?

8. What are the functions of the amnion, chorion and yolk sac in birds?

9. In what respects are the circulatory systems of the shark and the 48 hour chick embryo similar?

DAYS

1   Ovulation (secondary occyte inside zona pellucida, corona, and cumulus).

     Fertilization (oocyte short-lived; fertilizable for 24–36 hours).

2–4 Cleavages during oviducal passage, by cilia and peristalis of oviduct.

     Cumulus and corona cells removed by oviducal enzymes.

4–6 Blastocyst (blastula) free in uterus; differentiated into inner cell mass at one pole and outer trophoblast (extraembryonic ectoderm).

     Nourishment by uterine secretions, implantation-initiating factor from uterus dissolves zona pellucida and makes blastocyst sticky.

6–7 Implantation of blastocyst with pole at inner cell mass entering endometrium first; blastocyst invasive but endometrium must be receptive, mutual action.

     Delamination of entoderm from under surface of inner cell mass.

8   Future amniotic cavity arises by cavitation of inner cell mass.

     Entoderm extends ventrally to form a vesicle (forerunner of yolk sac).

     Plate of cells between two vesicles is future embryonic disk.

     Trophoblast differentiates into inner cytotrophoblast and outer syntrophoblast.

11–12 Extraembryonic mesoderm formed by delamination from inner cell mass and trophoblast into cavity of blastocyst.

     Trophoblastic villi (with mesodermal cores) penetrate deeper into endometrium; eroding blood vessels; nourishment now by blood (hemotrophic).

14   Extraembryonic coelom established by cavitation in extraembryonic mesoderm, giving a layer (somatic mesoderm) beneath trophoblast and over amniotic vesicle and a layer (splanchnic mesoderm) around entodermal vesicle. Fetal membranes now apparent, respectively: chorion (or serosa), amnion, yolk sac.

     Persisting bridge of mesoderm from embryonic disk to chorion is body stalk.

16   Formation of primitive streak on dorsal surface of embryonic disk.

     Establishment of embryonic mesoderm by involution through streak.

17   Allantois arises by tubular evagination of yolk sac into body stalk.

18–20 Head-fold, foregut, neurulation, etc. (as in chick).

     End of 2 months: annion and chorion fuse, obliterating extraembryonic coelom.

     End of 3 months: decidua capularis fuses with dicidua parietalis.

E. Development in the 72 Hour Chick Embryo

An exercise on the 72 hour chick whole mount has not been included because the increased thickness of the embryo, compared with the 48 hour embryo, makes observation of more than superficial structures difficult. Figure 49 will be helpful, however, in studying the cross sections. Note the advances made since the stage of 48 hours of incubation (compare figs. 39 and 48). Cranial and cervical flexures are sharper and caudal flexure of the tail-bud is apparent; the amnion completely encloses the embryo; the heart is more tightly coiled; an additional pharyngeal cleft and aortic arch have been added; allantois, wing and limb buds, and epiphysis are visible; and vitelline veins and arteries now parallel each other medially.

1. TRANSVERSE SECTIONS OF THE 72 HOUR CHICK EMBRYO: Follow the same general method of study used in the exercise on the 48 hour embryo (see p. 127). Examine the following structures.

Fetal membranes: Select any section passing through the anterior part of the embryo. Identify: *amnion, chorion,* and *yolk sac.* Their relationships are the same as those described for the 48 hour embryo (see p. 129). Is proamnion still present?

Nervous system and *special sense organs:*

Rhombencephalon: Owing to increased cranial flexure in the 72 hour embryo the anteriormost sections pass through the hindbrain, a large elongated vesicle seen in frontal aspect. Its large cavity is ventricle IV.

Myelencephalon: the posterior part of the hindbrain with thin sides and roof.

Metencephalon: the anterior region of the hindbrain with much thicker walls than those of the myelencephalon. There is no constriction in the brain delimiting metencephalon from myelencephalon.

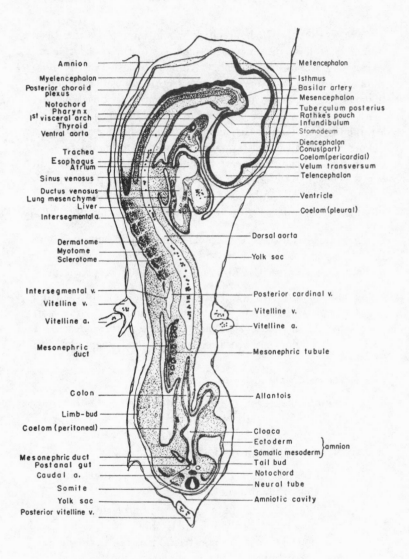

Fig. 48. Sagittal section of 72 hour chick embryo.

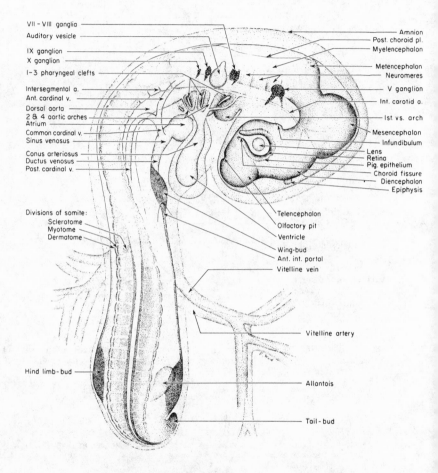

VII – VIII ganglia

Auditory vesicle

IX ganglion

X ganglion

I–3 pharyngeal clefts

Intersegmental a.

Ant. cardinal v.

Dorsal aorta

2 & 4 aortic arches

Atrium

Common cardinal v.

Sinus venosus

Conus arteriosus

Ductus venosus

Post. cardinal v.

Divisions of somite:

Sclerotome

Myotome

Dermatome

Hind limb-bud

Amnion

Post. choroid pl.

Myelencephalon

Metencephalon

Neuromeres

V ganglion

Int. carotid a.

Ist. vs. arch

Mesencephalon

Infundibulum

Lens

Retina

Pig. epithelium

Choroid fissure

Diencephalon

Epiphysis

Telencephalon

Olfactory pit

Ventricle

Wing-bud

Ant. int. portal

Vitelline vein

Vitelline artery

Allantois

Tail-bud

Fig. 49. Whole mount of 72 hour chick embryo.

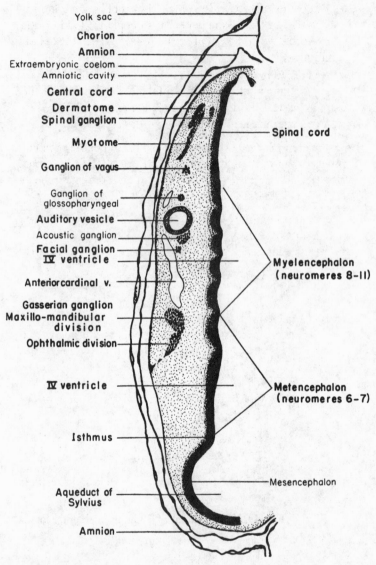

Fig. 50. Frontal section (cross section) through the brain of 72 hour chick embryo.

Posterior choroid plexus: the thin roof of the myelencephalon. In many specimens it is deeply folded into the myelocoel with, frequently, the amnion also. Vascularization occurs later.

Neuromeres: Observe that the embryonic segments or neuromeres of the neural tube are still evident in the walls of the hindbrain. The sixth neuromere, the first in the rhombencephalon, is very long.

Auditory vesicle: Disregarding the brain for the moment proceed caudally in the series until the auditory vesicles are located at the level of the 10th neuromere, one on each side of the myelencephalon.

Endolymphatic ducts: Differentiate between the endolymphatic duct, the small, thick-walled oval first encountered, and the larger auditory vesicle proper. Note the connection, if present, between the endolymphatic duct and the epidermis. Significance?

Utriculus (L. *utriculus* = diminutive of *uter*, meaning little bag): the more dorsal part of the auditory vesicle, possessing irregular outlines. Slight evaginations at the anterior and posterior ends of the utriculus mark beginnings of the *anterior* and *posterior oblique semicircular canals.* A faint bulge laterally foreshadows the *horizontal semicircular canal,* whereas the medial, thick-walled region is the utriculus proper.

Sacculus (L. *sacculus* = diminutive of *saccus,* meaning little sack): Continue in the series until the auditory vesicle becomes even in outline and thick-walled. This is the region of the future sacculus. Note its relationship to the auditory ganglion, to be studied later.

Mesencephalon: Return now to a section where the brain may appear like an exclamation point (!). The elongated part of the figure is the rhombencephalon described above, the circular region the mesencephalon with its cavity the mesocoel or future adqueduct of Sylvius. Trace caudally until these two regions fuse and their cavities become continuous. The constriction in the brain at this point is the isthmus (see fig. 48).

Cranial ganglia (associated with the hindbrain):

Acoustico-facialis ganglion of the *VII (facial)* and *VIII (auditory) nerves:* the dark mass of cells anterior to the auditory vesicle. Both facial and auditory elements are included. Examine under high power a dorsal section showing the nerve fibers passing between hindbrain and ganglion. Trace ventrally (i.e., caudally in the series). At the level of the sacculus the auditory and facial parts can be distinguished, the former, noted in connection with ear, as a mass in intimate contact with the sacculus and the latter as a division proceeding in an anterolateral direction toward the skin (to be traced later).

Gasserian (semilunar) ganglion of the *V (trigeminal) nerve:* large dark mass anterior to the acoustico-facialis ganglion. Observe closely its relationship to the brain (neuromere 7) and its size and shape. Trace it laterally. Note small twigs (nerves) passing to the skin. Is the epidermis specialized at these points? Farther ventrally the trigeminal divides into an anterior ramus, the *ophthalmic* nerve, and a posterior division which later gives rise to the *mandibular* and *maxillary* nerves.

Ganglion of the *IX (glossopharyngeal) nerve:* small ganglionic mass lying posterior to the auditory vesicle and adjacent to the myelencephalon.

Ganglion of the *X (vagus) nerve:* still smaller ganglionic mass, sometimes not yet separated from the glossopharyngeal ganglion, lying posterior to the IX nerve.

Third cranial nerve (oculomotor): Continue caudally in the series from the level just studied until the neural tube "breaks" into two segments: the mesencephalon and the spinal cord now flanked by the first pair of somites, long thin plates to be studied later. Trace farther. The *notochord* will be observed in passing. Soon the oculomotor nerves will appear as a pair of streaks emerging from the floor of the mesencephalon. Details of their distribution will be considered later.

Diencephalon: Disregarding other systems for the time, continue posteriorly in the series until the rotund, thick-walled mesencephalon is replaced by the laterally compressed, somewhat thinner-walled diencephalon. Examine under high power. Compare inner and outer layers of the diencephalic wall. Where are mitotic figures found?

Infundibulum: the long ventral outpocketing of the diencephalon.

Eye:

Retina: the inner layer of the optic cup. Examine under high power. Is the arrangement of the layers similar to that in the brain? Where are mitotic figures most abundant? What will these dividing cells become?

Pigmented epithelium: the outer layer of the optic cup, not yet pigmented.

Lens vesicle: Examine under high power. Note the shape of the central cells and the position of their nuclei. What will these cells form? What becomes of the outer narrower layer of cells? the lens cavity?

Mesodermal coats: Note beginning condensations of mesenchyme outside the optic cup to form a layer which will later differentiate into the *choroid* and the *sclera.*

Choroid fissure: Find a section in which the ventral lip of the optic cup is absent or seems to be perforated. This is owing to the choroid fissure, a deep cleft in the ventral rim of the cup. Via this fissure a blood vessel, the central artery, may be seen entering the eye.

153

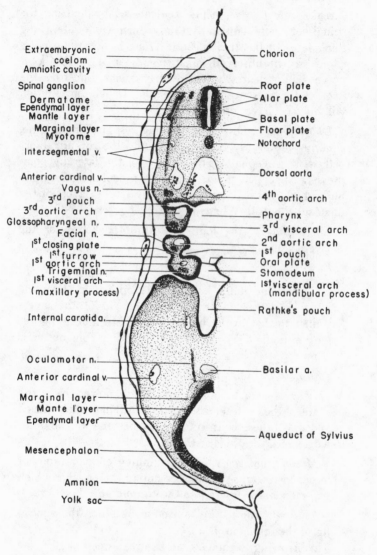

Extraembryonic coelom
Amniotic cavity
Spinal ganglion
Dermatome
Ependymal layer
Mantle layer
Marginal layer
Myotome
Intersegmental v.
Anterior cardinal v.
Vagus n.
3rd pouch
3rd aortic arch
Glossopharyngeal n.
Facial n.
1st closing plate
1st furrow
1st aortic arch
Trigeminal n.
1st visceral arch
(maxillary process)
Internal carotid a.
Oculomotor n.
Anterior cardinal v.
Marginal layer
Mantle layer
Ependymal layer
Mesencephalon
Amnion
Yolk sac

Chorion
Roof plate
Alar plate
Basal plate
Floor plate
Notochord
Dorsal aorta
4th aortic arch
Pharynx
3rd visceral arch
2nd aortic arch
1st pouch
Oral plate
Stomodeum
1st visceral arch
(mandibular process)
Rathke's pouch
Basilar a.
Aqueduct of Sylvius

Fig. 51. Cross section of 72 hour chick embryo at level of third cranial nerve. Pharyngeal region cut frontally.

154

Optic stalk: the hollow bridge connecting the optic cup and diencephalon. Is the third ventricle of the diencephalon continuous with the cavity of the optic cup or with the remnants of the cavity of the optic vesicle?

Telencephalon: Proceed caudally in the series (actually cranially in the forebrain) until the brain begins to dilate laterally in the ventral part. The region of lateral expansion is now telencephalon and the outpocketings, bearing ventricles I and II, are the forerunners of the cerebral hemispheres. Dorsally, however, one still sees diencephalon which overhangs the cerebrum.

Epiphysis: a mid-dorsal evagination of the diencephalon. Is it directed anteriorly or posteriorly?

Lamina terminalis: the anteriormost wall of the neural tube in the midline, seen here as a thin section of the anterior telencephalic wall; the bulging hemispheres will later extend forward beyond the lamina terminalis.

Nasal pits: ectodermal invaginations to each side of the telencephalon. The openings to the outside (amniotic cavity) will become the external nares.

Spinal cord: Examine the spinal cord at the opposite end of the sections just studied.

Central canal: the neurocoel in the spinal cord region of the neural tube.

Floor plate: the narrow mid-ventral wall of the spinal cord.

Roof plate: the mid-dorsal wall of the spinal cord.

Sulcus limitans: lateral depression of the central canal, almost imperceptible at this stage, about midway between floor and roof plates.

Alar plates: the dorsolateral wings of the spinal cord, i.e., the wall between sulcus limitans and roof plate.

Basal plates: the ventrolateral wings of the spinal

155

cord, i.e. the wall between sulcus limitans and floor plate.

Marginal layer: the outer or peripheral region of the neural tube characterized by many fibers (nervous and supportive) and fewer nuclei.

Ependymal layer: the innermost layer of cells adjacent to the central canal, usually exhibiting mitotic figures. Examine under high power.

Mantle layer: the wide heavily nucleated layer between marginal and ependymal layers.

Dorsal (sensory) root ganglia: Best seen farther forward in the body; return, therefore, to sections passing through the diencephalic regions of the brain. Note the dorsal root ganglia as small loosely organized dark bodies, one dorsolaterally situated on each side of the spinal cord.

Ventral (motor) roots: in favorable specimens, faint paired twigs composed of nerve fibers and nuclei will be observed emerging from the ventrolateral walls of the spinal cord and passing but a short distance in the adjacent mesenchyme.

Foregut and associated structures (paired structures described as single):

First pharyngeal furrow and *pouch:* Find the section in which the oculomotor nerves leave the mesencephalon. Observe the first pharyngeal furrow as a V-shaped ectodermal indentation. The furrow opens directly into the pharyngeal region of the foregut via an entodermal evagination, the first pharyngeal pouch. Is the cleft very long vertically?

First visceral (mandibular) arch: the mass of mesenchyme anterior to the first cleft (i.e., toward the mesencephalon in the section).

Second pharyngeal furrow and *pouch:* The second pharyngeal unit, similar to number one, appears more posteriorly, i.e., toward the spinal cord in the section.

Second visceral (hyoid) arch: the mesenchyme between the first (hyomandibular) pouch and furrow and the second cleft.

Stomodeum: The first or mandibular arch seems to be cut away from the head by a deep "cleft" which opens into a medial chamber, the stomodeum.

Oral plate or *pharyngeal membrane:* a very thin membrane between the two halves of the mandibular arch; sometimes broken in the 72 hour embryo.

Rathke's pouch: an irregular cavity lined with cuboidal epithelium anterior to the stomodeum with which it is confluent. Study the relationship of Rathke's pouch to the infundibulum.

Third pharyngeal furrow and *pouch:* A third pharyngeal unit, similar to the first two, appears still more posteriorly in the frontal sections of the foregut. Is the closing plate ruptured?

Symphysis of the lower jaw: formed by the ventral fusion of the halves of the mandibular arch (mandibular processes).

Maxillary process: a lobe-like mass of mesenchyme, covered with epidermis, situated between the eye and the mandibular process. The maxillary processes form the greater part of the upper jaws.

Thyroid: a diverticulum extending ventrally toward the lower jaw from the floor of the pharynx at the level of the second pair of pouches. Examine its epithelium under high power.

Fourth pharyngeal furrow and *pouch:* The series of pharyngeal units is finally completed by the fourth furrow and pouch posterior to the others. Is a cleft present?

Laryngotracheal groove: Posterior to the fourth pair of

Pharyngeal Derivatives in Higher Vertebrates

Unit number	Visceral arches[1]	Pharyngeal pouches	Aortic arches
I	Jaws (mandibular and maxillary components) Malleus, Incus	Middle ear Eustachian tube Auditory canal (from first furrow)	——
II	Hyoid apparatus Stapes	Palatine tonsil[2]	——
III	Hyoid apparatus	Inferior parathyroid Thymus	Common carotid artery Base of internal carotid artery[3]
IV	Thyroid cartilage	Superior parathyroid Thymus	Part of arch of aorta[4] Part of right (mammal) or left (bird) subclavian artery
V	Cricoid and arytenoid cartilages	Ultimobranchial body (lateral thyroid)	
VI	——	——	Proximal part of pulmonary artery Ligamentum arteriosum (see p. 236)

[1] The first visceral arch is the mandibular arch, the second one the hyoid arch, from three on the branchial arches, in gill-bearing vertebrates.

[2] The thyroid gland arises from the floor of the pharynx at the level of the first or second pouch.

[3] The remainder of an internal carotid comes from a dorsal aorta (anterior to third arch) and an extension thereof into the head. An external carotid arises from segments of the ventral aorta (anterior to third arch) and extensions thereof into the lower jaw.

[4] Other parts of the adult aorta come from the ventral aorta (aortic sac) and an embryonic dorsal aorta posterior to the fourth arch.

pharyngeal pouches the foregut becomes elongated ventrally. The ventral-most V-shaped part is the laryngotracheal groove.

Esophagus: the dorsal somewhat rounded part of the foregut unseparated, at this level, from laryngotracheal groove.

Lung buds: pair of lateral diverticula from the posterior end of the laryngotracheal groove, now separated from the esophagus above. Trace each lung bud posterolaterally. Technically, the circular "doughnuts" just cut off from the laryngotracheal groove are the future *bronchi.* They lead to the lung buds proper, thick-walled ovals at the posterior ends of the bronchi.

Stomach: Trace the esophagus posteriorly until the entodermal tube enlarges. This region of the foregut is future stomach.

Duodenum: the region of the foregut posterior to the stomach in which the entodermal tube is a vertically oriented oval.

Anterior (dorsal) liver diverticulum: an elongated diverticulum situated between a large blood channel, ductus venosus (to be studied later), and the duodenum with which the liver diverticulum is connected as may be demonstrated by tracing. The anterior liver diverticulum extends toward the right side of the embryo (away from the yolk-sac) along the wall of the ductus venosus.

Posterior (ventral) liver diverticulum: Follow the duodenum as it moves down along the left side of the ductus venosus. Observe another liver diverticulum similar to the one just described, lying ventral to the ductus venosus and connecting to the duodenum. This is the posterior diverticulum which gives liver buds to right and left where they are embedded in the mesenchyme surrounding the ductus venosus.

Anterior intestinal portal: Follow the duodenum as it

moves to a position ventral to the ductus venosus. It curves then to the left and opens onto the yolk via the familiar anterior intestinal portal.

Circulatory system (paired structures described as single):

Anterior cardinal vein: Find this large vein laterally situated in the mesenchyme of sections passing through the floor of the hindbrain. Note the relation of the vein to the acoustico-facialis and gasserian ganglia. The facial nerve and first pharyngeal furrow seem to split the vein into an anterior segment which can be traced forward in the head and a posterior segment which will be followed caudally.

Internal carotid artery: a narrower vessel, usually containing blood cells, medial to the anterior cardinal vein just observed and running in the mesenchyme lateral to the notochord. The first pharyngeal pouch seems to break this vessel into two segments. Trace the anterior segment, as internal carotid, forward around Rathke's pouch and up along the sides of the infundibulum and diencephalon. Observe that the two internal carotids are connected by a delicate plexus posterior to the infundibulum (i.e., between infundibulum and mesencephalon). This cross trunk is part of the *circle of Willis* (after Thomas Willis, English anatomist, 1621-1675). If time is available later in the exercise the more anterior distribution of the internal carotids may be studied.

First aortic arch: Return to the section described above in which the first pharyngeal pouch seems to break the internal carotid. Note that the anterior segment just traced immediately gives off (actually receives) a very small vessel which can be traced ventrally, in favorable specimens, in the substance of the mandibular arch. This is the first aortic arch.

Second aortic arch: Trace now the posterior segment of the internal carotid artery which shall hereafter be designated the paired dorsal aorta; note that a vessel larger than the first aortic arch is received from the second visceral arch. This is the second aortic arch.

Third aortic arch: In like manner a third still larger vessel is received by the paired dorsal aorta from the third visceral arch.

Unpaired dorsal aorta: Posterior to the third pharyngeal pouches observe that the paired dorsal aortae fuse to form the unpaired dorsal aorta.

Fourth aortic arch: Shortly posterior to the point of fusion of the paired dorsal aortae the fourth aortic arch emerges from the substance of the fourth visceral arch.

Sixth aortic arch: a smaller vessel entering the fourth arch from the mesenchyme posterior to the fourth pharyngeal pouch. A fifth aortic arch can not be seen at this stage. In favorable specimens the sixth arch can be traced ventrally.

Ventral aorta: Move back in the series of sections to a point where the first aortic arches may be observed as separate vessels in the mandibular processes (arch). Proceed again caudally in the series noting that the first aortic arches "fuse" just anterior to the thyroid to "form" the ventral aorta. Almost immediately the second and third arches "join" the ventral aorta. At the level of "entrance" of the fourth arches the ventral aorta may be observed to be continuous with the anterior end of the conus of the heart.

Intersegmental arteries and *veins:* Note small vessels (intersegmental arteries) passing dorsally from the dorsal aorta and others (intersegmental veins) entering the anterior cardinal veins. These vessels supply the dorsal organs such as somites, neural tube, notochord, etc.

Pulmonary artery: In some preparations a small vessel will emerge from the base of the sixth arch, just as the latter enters the base of the fourth arch or ventral aorta. This vessel, the pulmonary artery, runs in the mesenchyme

161

to the side of the laryngotracheal groove. It can be traced for some distance in some specimens toward the lung bud.

Conus: The ventral aorta gives way, as we have seen, to the conus of the heart which is swung for a short distance by a *dorsal mesocardium* and which is enclosed by the pericardial cavity. The conus swings to the right.

Atrium: The atrium comes into view as one proceeds caudally in the series as a large chamber to the left of the conus. Why are the blood corpuscles piled up against one wall instead of being evenly distributed throughout the cavity of the chamber? Study the epimyocardium under high power. Is it differentiating into its two tissues? What is the relation of the endocardium to the epimyocardium?

Sinus venosus: At the level of the posterior end of the lung buds, the dorsal part of the left-hand chamber, designated atrium above, becomes set apart by a slight constriction as a separate chamber, the sinus venosus, which is supported by a dorsal mesocardium.

Common cardinal or *duct of Cuvier:* a continuation of the anterior cardinal vein ventrally in the lateral body wall. Observe its entrance into the sinus venosus.

Ductus venosus: Tracing the sinus venosus posteriorly note that it becomes a large vessel, the ductus venosus, which is surrounded by a thick wall of mesenchyme in which are embedded the liver diverticula.

Ventricle: By now the right and left sections of the heart have fused and we are clearly in the ventricular region. Sharp boundaries between conus and ventricle and between ventricle and atrium do not exist.

Trabeculae: irregular projections of the myocardium into the cavities of the heart, especially numerous in the wall of the ventricle. Note that the endocardium dips into the spaces between the trabeculae. The cardiac muscle forms a compact layer distally, just under the epicardium.

Vitelline (omphalomesenteric) veins: Continue caudally in the series. Note that the ductus venosus becomes "di-

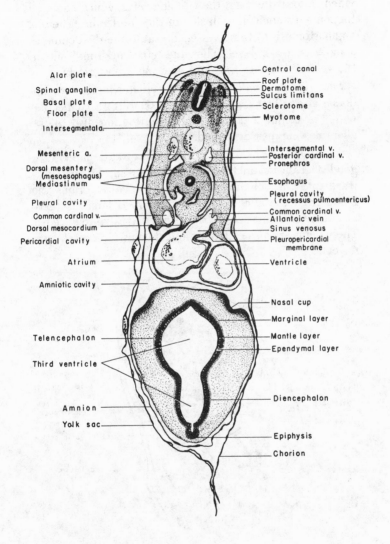

Fig. 52. Cross section of 72 hour chick embryo at level of heart.

Alar plate
Spinal ganglion
Basal plate
Floor plate
Intersegmental a.
Mesenteric a.
Dorsal mesentery (mesoesophagus)
Mediastinum
Pleural cavity
Common cardinal v.
Dorsal mesocardium
Pericardial cavity
Atrium
Amniotic cavity
Telencephalon
Third ventricle
Amnion
Yolk sac

Central canal
Roof plate
Dermatome
Sulcus limitans
Sclerotome
Myotome
Intersegmental v.
Posterior cardinal v.
Pronephros
Esophagus
Pleural cavity (recessus pulmoentericus)
Common cardinal v.
Allantoic vein
Sinus venosus
Pleuropericardial membrane
Ventricle
Nasal cup
Marginal layer
Mantle layer
Ependymal layer
Diencephalon
Epiphysis
Chorion

vided" into two vessels, the vitelline veins, which continue for some distance on each side of the duodenum. There is an anastomosis of these vessels dorsal to the duodenum at a more posterior level. Then the left vein "runs" out on the yolk-sac as does the right vein more posteriorly.

Posterior cardinal: a vessel immediately lateral to the dorsal aorta. Trace it anteriorly to establish a connection with the common cardinal, giving attention also to the allantoic vein next described.

Allantoic vein (G. *allantoeides*=sausage-shaped): a good-sized vessel in the lateral body wall. Trace it forward into the common cardinal. This vessel drains the blood from the allantois, to be described later. Is there any difference in size of right and left allantoic veins? These veins are also called umbilical veins (L. *umbilicus*= navel).

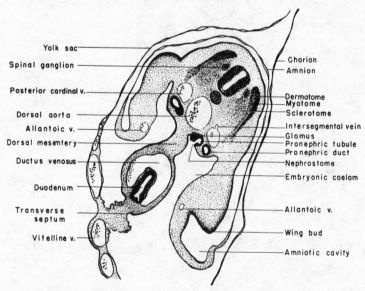

Fig. 53. Cross section of 72 hour chick embryo just cranial to anterior intestinal portal.

Primary derivatives	Secondary derivatives or other fate	Function
I. Ventral mesentery		
1. Region of heart	Dorsal and ventral mesocardia; fleeting existence	Support heart in early stages
2. Region of liver	Gastro-hepatic ligament‡	Connects stomach and liver
	Hepato-duodenal ligament‡ (‡ together form lesser omentum)	Connects liver and duodenum; supports bile duct
	Falciform ligament	Connects liver to ventral body wall
3. Region of esophagus	Ventral part of mediastinum	Supports trachea
II. Dorsal mesentery		
1. Mesoesophagus	Dorsal part of mediastinum	Supports eosphagus and other organs in mediastinum
2. Mesogaster	Gastro-splenic ligament	Connects stomach and spleen
	Spleno-renal ligament	Connects spleen and left kidney
	Gastro-colic ligament	Connects stomach and transverse colon (result of fusion with colon)
	Greater omentum	Vascularized apron over intestines; fat storage
3. Mesoduodenum	Disappears	Duodenum fused to body wall
4. To cranial limb of intestinal loop	Mesentery proper	Connects jejunum and ileum to body wall
5. To caudal limb of intestinal loop	Ascending mesocolon (disappears)	Ascending colon fused to body wall
	Transverse mesocolon	Swings transverse colon
	Descending mesocolon (disappears)	Descending colon fused to body wall
	Sigmoid mesocolon	Supports sigmoid colon
6. Mesorectum	Disappears	Rectum fused to body wall

[1] This table is based primarily upon mammalian development.

Partitioning of the coelom:

Dorsal mesentery: the mesoderm supporting the gut from the dorsal body wall and dividing the body cavity dorsally into right and left compartments; seen to advantage in sections immediately anterior to the anterior intestinal portal.

Ventral mesentery: a midline partition of mesoderm ventral to the gut; present only in certain limited regions and then incompletely. Two examples:

Find a section showing the ductus venosus and the anterior and posterior liver diverticula. The isthmus of mesoderm between duodenum and the structures just named is a part of the ventral mesentery, as also the loose mesenchyme in the midline ventral to the ductus venosus and the posterior liver diverticulum. This part of the ventral mesentery is the forerunner of the *falciform ligament*.

Technically speaking the *dorsal* and *ventral mesocardia* are specializations of the ventral mesentery because they constitute a midline partition ventral to the gut. The ventral mesocardium has a very fleeting existence and the dorsal mesocardium does not persist long.

Pleuropericardial membrane (lateral mesocardium): Return to the section showing the duct of Cuvier entering the sinus venosus. The bridge of mesenchyme, through which the duct of Cuvier passes, is the pleuropericardial membrane which separates in a limited region the pleural and pericardial cavities.

Pleural cavities: the subdivision of the coelom noted above into which the lungs grow. Each cavity appears as a long irregular vertically oriented cavity.

Pericardial cavity: the part of the coelom immediately surrounding the heart and separated, as just noted, from the pleural cavities by the pleuropericardial membranes.

This separation is not complete yet; the two cavities are continuous anterior and posterior to the narrow membranes.

Mediastinum (L. *mediastinus* = medial): the name given to the whole midline partition in the region of the future thorax. It includes dorsal mesentery (specifically mesoesophagus) and ventral mesentery (dorsal mesocardium).

Transverse septum: Several sections posterior to the entrance of the common cardinals into the sinus venosus find a section through the ductus venosus and the branching liver buds. The mesenchyme surrounding these structures forms an incomplete partition transversely across the embryo. This is the transverse septum which is not readily distinguished from the pleuropericardial membranes and the ventral mesentery. Perhaps the simplest picture is that of one quite thick partition at the level of the sinus venosus and ductus venosus to which the following contribute: 1) pleuropericardial membranes—mesenchyme about the ducts of Cuvier, 2) ventral mesentery—mesenchyme in the midline, and 3) transverse septum—mesenchyme lateral to the ductus venosus.

Peritoneal cavity: the body cavity posterior to the transverse septum which is continuous with both the pleural and pericardial cavities at this stage.

Urogenital system:

Mesonephric ridge: the rounded bulge of the dorsal body wall into the peritoneal cavity, containing the mesonephros.

Mesonephros:

Mesonephric duct or *Wolffian duct* (after Kaspar Friedrich Wolff, German anatomist and embryologist, 1733–1794): thick-walled tube situated far laterally,

167

just under the peritoneum and ventral to the posterior cardinal vein.

Mesonephric tubules: paired S-shaped tubules medial to the mesonephric duct, with which they are connected. Study a number of tubules at different levels of the body in search of one which will show some evidence of the connection between tubule and duct and of blood vessels between tubule and the dorsal aorta and posterior cardinal veins. Note also a small vessel ventral to the Wolffian duct; this is the *subcardinal vein.*

Genital ridge: a slight bulge, seen in some sections, immediately lateral to the dorsal mesentery of the midgut but medial to the mesonephric ridge. Mesonephric and genital ridges together may be designated the *urogenital ridge.*

Pronephros: Cranial to the anterior intestinal portal may be seen vestiges of the pronephros in the form of diminutive tubes or small bodies ventrolateral to the dorsal aorta. Nephrostomes may be seen, or only notches in the coelomic epithelium where nephrostomes did occur. The region of the pronephros is given as somite levels 5 to 15 or 16, the region of the mesonephros as somite levels 13 to 30. Review the pronephros of the frog larva (p. 80) and reread the discussion on the vertebrate nephric system (p. 142).

Structures at the level of the hindgut:

Hindgut: the posterior region of the gut which, like the foregut, possesses a floor.

Posterior intestinal portal: the boundary between the hindgut and midgut at which the hindgut opens onto the yolk.

Allantois: To the right of the hindgut and lying in the body cavity, observe a dark mass containing a large irregular sacculation. This is one of the fetal membranes,

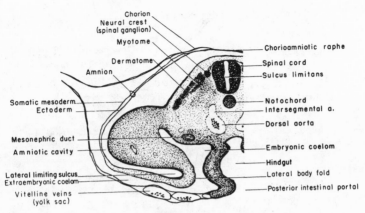

Fig. 54. Cross section of 72 hour chick embryo at level of posterior intestinal portal.

the allantois, consisting of an entodermal lining and a mesodermal covering (the mass of mesenchyme enclosing the entoderm). Trace posteriorly to demonstrate the connection between allantois and hindgut.

Cloaca: the region of the hindgut into which open the allantois and mesonephric ducts. Find the section showing the connection of the ducts to the cloaca. Does the connection appear to be functional?

Subcaudal pocket: Continue posteriorly until the embryo is cut free of the blastoderm. The space between embryo and blastoderm is the subcaudal pocket, the consequence of the *tail fold* undercutting the posterior tip of the embryo as did head fold in establishing the subcephalic pocket.

Cloacal membrane or *anal plate:* Observe that the narrow floor of the cloaca is in close contact with the ventral ectoderm. This plate of entoderm and ectoderm is the cloacal membrane which upon rupture will establish the anal opening (cf. oral plate or pharyngeal membrane).

Tail: At about the level of the anal plate there appears

169

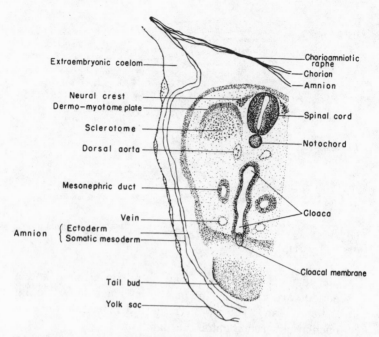

Fig. 55. Cross section of 72 hour chick embryo at level of cloaca.

a rounded or oval mass of tissue to the right of the embryo. This is the tip of the tail, which because of *caudal flexure* is curved forward and is being cut transversely at this level. Proceed caudally to demonstrate the connection of tail and embryo proper.

Torsion: Note that the posterior end of the embryo is twisting, like the head, so that the embryo comes to lie on its left side.

Postanal gut: the blind part of the hindgut caudal to the anal plate. Observe that it extends as a long narrow tube into the tail (see fig. 56). It will later disappear.

Tail bud: large mass of indifferent cells, the remnants of the primitive streak and Hensen's node, situated subterminal to the tip of the tail (see fig. 56).

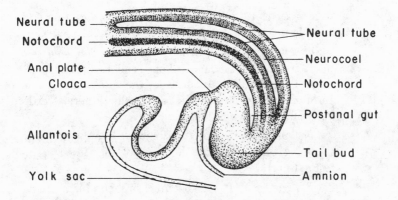

Neural tube
Notochord
Anal plate
Cloaca
Allantois
Yolk sac

Neural tube
Neurocoel
Notochord
Postanal gut
Tail bud
Amnion

Fig. 56. Sagittal section through posterior part of 72 hour chick embryo.

Other structures: Examine the posterior ends of the notochord and neural tube. What type of section through the body is here exemplified?

Miscellaneous structures:

Somite: Select for study a somite from the level of the diencephalon. Examine under high power.

Sclerotome: the loosely arranged, dark group of cells lateral to the neural tube and notochord.

Myotome: a long light stripe lateral to the sclerotome and extending ventrally from the dorsal rim of the somite like a tail toward the posterior cardinal vein.

Dermatome: the dark layer of cells forming the outermost part of the somite and lying lateral to the dorsal part of the myotome.

Limb buds:

Wing buds: At a level caudal to the anterior intestinal portal the wing buds may be seen as prominent rounded or conical bulges on each side of the body; each con-

171

sists of a covering of thickened ectoderm and a core of mesoderm (somatic mesoderm of the lateral plate).

*Subclavian artery** (L. *sub* = under + *clavicula* = a small crossbar, i.e., clavicle bone): in favorable specimens a small vessel arising from the side of the dorsal aorta and proceeding laterad above the posterior cardinal vein into the wing bud.

Leg buds: paired structures, similar in appearance to the wing buds, lying at the level of the posterior intestinal portal.

*Iliac arteries** (L. *ilia* = loins): in favorable preparations, small branches from the dorsal aorta which may be traced into the leg buds.

Notochord: An examination of this structure under high power is worthwhile in that it shows the beginning vacuolization of the notochordal cells. The stiffness of the notochord, like that of a plant, is owing to the turgidity of these vacuoles which later come to occupy almost all of the interior of the cell.

Pulmonary veins: small irregular vessels lying ventral to the bronchi and passing by a common vessel via the dorsal mesocardium into the left side of the atrium just to the left of the anterior part of the sinus venosus.

Basilar artery: a median unpaired vessel, best seen ventral to the mesencephalon. Trace it forward into the internal carotids (circle of Willis) and posteriorly under the brain as far as possible.

Mesenteric arteries: small, unpaired, ventral branches of the dorsal aorta passing in the dorsal mesentery to the gut. Examine the levels of the embryo from laryngotracheal groove to the anterior intestinal portal for examples of these vessels.

Caudal arteries: continuations of the dorsal aortae into the tail.

Allantoic or *umbilical arteries** (may not be present):
a branch, one on each side of the body, of the iliac artery
leading to the allantois.

External carotid arteries: a pair of small vessels extending forward from the bases of the first aortic arches into
the lower jaw (mandibular processes).

Pancreas:* Find the section of the embryo in which the
ventral liver diverticulum leaves the duodenum. The pancreas is represented, if present, by a thick diverticulum
from the dorsal wall of the duodenum.

Finer details in the distribution of the *cranial nerves* (most
of these features will be illustrated to better advantage in the
serial sections of the 10 mm. pig embryo):

Olfactory nerve (I): The beginnings of this nerve
(paired) may be seen by tracing a small strand of nuclei
and fibers from the anteromedial wall of the nasal pit
forward to the lateral wall of the telencephalon.

Oculomotor nerve (III): The root of this nerve (paired)
as it leaves the ventral wall of the mesencephalon has
been identified earlier. Trace the nerve ventrally. It is
shortly lost in the mesenchyme. Later the third cranial
nerve passes forward to innervate four of the extrinsic
muscles of the eye (superior, inferior, and internal rectus
and inferior oblique). There will develop also in association with the oculomotor autonomic nerve fibers which
supply the intrinsic (smooth) muscles of the eye.

Trigeminal nerve (V): Trace the ophthalmic and maxillomandibular branches as far as possible. The ophthalmic branch may be followed in favorable specimens
through the mesenchyme between the dorsal rim of the
optic cup and the brain into the anterior part of the head.
The maxillomandibular branch may be traced ventrally
into the first arch with the greater part passing into the
mandibular process and a smaller branch (not well
shown) into the maxilliary process.

Facial nerve (VII): Follow the facial nerve noting its relation to the acoustico-facial ganglion, the anterior cardinal vein, and finally the first or hyomandibular cleft. A small branch (pretrematic) runs down the anterior surface of the cleft (hence posterior side of the mandibular arch) whereas the larger branch (posttrematic) passes into the hyoid arch.

Glossopharyngeal nerve (IX): Trace this nerve from its ganglion to its termination in the third visceral arch.

Vagus nerve (X): can not be traced at this stage. Parts of it, however, may be seen in the dorsal part of the fourth visceral arch.

Spinal accessory nerve (XI): represented by a longitudinal bundle of fibers and nuclei extending back toward the spinal cord along the dorsolateral wall of the posterior end of the myelencephalon; readily seen at the level of the first somite.

Hypoglossal nerve (XII): not well shown; represented by large dots connected to the ventrolateral wall of the myelencephalon near its posterior boundary.

Review such features, described in the exercise on the 48 hour chick, as: *lateral body folds, lateral limiting sulcus, chorioamniotic raphe, extraembryonic coelom.*

Review Questions

1. With what regions of the brain are the twelve cranial nerves associated?
2. What is the developmental origin of: olfactory nerve, optic nerve, oculomotor nerve, cranial ganglia?
3. Name the structures supplied by cranial nerves III, IV, and VI. What relation do cranial nerves V, VII, IX, and X have to the visceral arches?
4. Trace the blood through the heart of the adult bird. In

what respects is the adult heart unlike that of the 72 hour chick?

5. Give the arterial supply and venous drainage of the following: myelencephalon, spinal cord (posterior level), diencephalon, lung buds, allantois, leg buds.

6. Draw in color the fetal membranes of the chick. What are the functions of the allantois? How do the fetal membranes of the chick differ, developmentally and functionally, from the fetal membranes of the mammal?

7. Define: visceral peritoneum, parietal pleura, pericardium, mesentery, omentum, falciform ligament, mesocardium, transverse septum, pleuropericardial membranes, ductus venosus, glomerulus, genital ridge, osteoblast, scleroblast, chondroblast, myoblast, erythroblast, odontoblast, dermoblast.

GENESIS OF NERVE FIBERS

Prior to 1910 the origin of nerve fibers was an unsolved problem, but Ross Harrison (see back cover of manual) proved in the first use of cell culture that axons and dendrites are outgrowths of embryonic nerve cells (neuroblasts). Most axons (parts of nerve cells that carry action potentials—dendrites are generating fibers) are myelinated (or medullated), that is, they possess a lipid-rich myelin sheath. Notable exceptions are postganglionic axons of the autonomic system, a part of an axon adjacent to a nerve cell body, and axon terminals. Electron miscroscopy elucidated the origin of a myelin sheath. In the peripheral nervous system sheaths are formed by sheath cells (of neural crest origin), wrapping their cell membranes around a naked axon until a thick, multilaminar covering is produced. Each sheath cell forms a segment of myelin; the junctions between segments are Nodes of Ranvier. An impulse jumps from node to node. Accordingly, myelinated fibers transmit messages rapidly. In the central nervous system, myelin formation (in white matter only) is similarly created by oligodendroglia—part of the connective tissue of brain and spinal cord. Arms of one glial cell, however, can myelinate segments of several nerve fibers.

175

V

DEVELOPMENT IN THE PIG EMBRYO AND FETUS

A. Transverse Sections of the 10 MM. Pig Embryo

The 10 mm. pig embryo has been selected for the exercise following the 72 hour chick because it illustrates a more advanced stage in the development of the vertebrate body. Moreover, the picture is that of the mammal which in later stages becomes increasingly different from patterns of development in other vertebrates. The general similarity of the 10 mm. pig embryo to the 72 hour chick with respect to curvatures of the body, basic plans in the nervous, digestive, and circulatory systems, and the plane of section in the series makes orientation easy. The procedure of tracing a single system, temporarily ignoring others, will be used here as in the preceding exercise.

Divisions of the brain: Owing to cranial flexure the first sections will pass frontally through the hindbrain as in the 72 hour chick embryo. Before studying details it is desirable that the five major divisions of the brain be identified by examining the serial sections in large jumps, that is, by rows. Note the following.

Myelencephalon: the part of the hindbrain with large wide ventricle (IV) and thin dorsolateral walls at anterior levels but exhibiting a narrower cavity and thicker walls posteriorly.

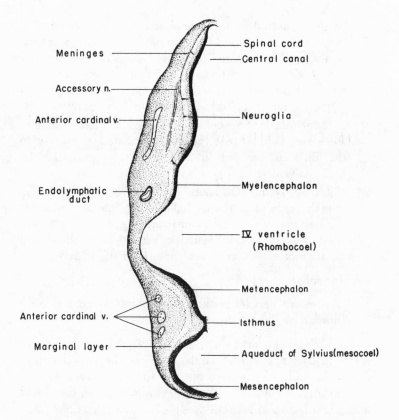

Fig. 57. Frontal section (cross section) through brain of 10 mm. pig embryo.

Metencephalon: the anterior part of the hindbrain, seen in the frontal sections as a region with thick walls and large V-shaped cavity (anterior part of ventricle IV).

Mesencephalon: a rounded or oval thick-walled region of the brain anterior to the metencephalon. The constriction, or more properly the connection, between mesencephalon and metencephalon is the familiar *isthmus*. Continue caudally in the series until mesencephalon be-

177

comes separated from the hindbrain, and rhombencephalon in turn is separate from spinal cord. Thus, a frontal section will show three parts of the central nervous system. Assistance in visualizing the plane of section may be obtained by using a straightedge on a text figure of a whole mount or sagittal section of the 10 mm. pig.

Diencephalon: Proceeding to the third or fourth slide in the series observe that the mesencephalon gives way to the diencephalon which is laterally compressed. The cavity is ventricle III.

Telencephalon: On more posterior slides identify the laterally expanded telencephalon. Each lateral outpocketing becomes a cerebral hemisphere; the cavities are ventricles I and II (lateral ventricles) in the hemispheres and an anterior extension of ventricle III in the midline.

Special sense organs:

Auditory vesicle: Return now to the sections passing through the hindbrain and the auditory vesicles.

Endolymphatic duct: small, irregular oval lateral to myelencephalon. Trace the duct into the medial wall of the auditory vesicle proper. Is there any connection of the duct with the embryonic epidermis as in the chick?

Anterior and *posterior oblique* and *horizontal semicircular canals:* slight evaginations from the utricular part of the auditory vesicle as described for the 72 hour chick embryo (see p. 151).

Sacculus: the ventralmost part of the auditory vesicle which is in close relationship with the auditory nerve.

Eye: Study sections through the diencephalon showing the optic cups or future eyes to each side of the brain.

Retina: the inner layer of the optic cup. Examine the retina under high power. The nuclei have not yet become organized into the definitive layers seen in the

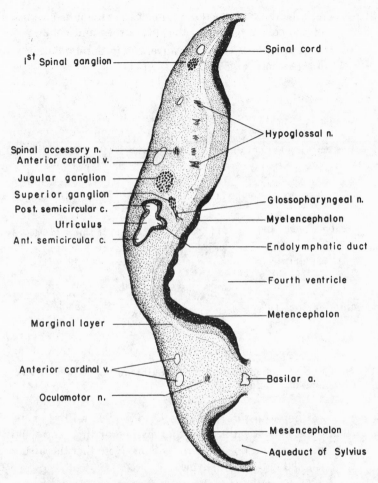

Fig. 58. Frontal section (cross section) through brain of 10 mm.
pig embryo, ventral to that shown in figure 44.

retina of the amphibian larva studied previously. Note
fine nerve fibers emerging from the inner surface of the
retina; these are the axones of the future ganglion cells.

Pigmented epithelium: the outer layer of the optic

179

cup. Note that brownish pigment is forming in the ends of the cells adjacent to the future rods and cones.

Lens vesicle: not so far advanced in differentiation as the lens in the 72 hour chick embryo.

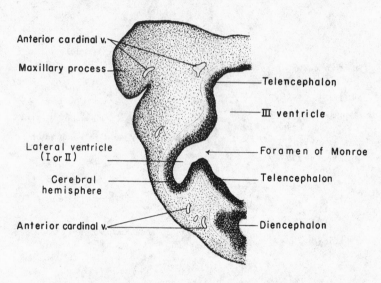

Fig. 59. Cross section through head of 10 mm. pig embryo at level of foramina of Monroe.

Choroid fissure: a deep fold in the ventral rim of the optic cup, identified by the absence of the ventral lip of the cup in one or more sections. Note that the fissure is filled with mesenchyme.

Optic stalk: the patent connection of the optic cup to the diencephalon.

Cornea: the outer covering of the eye, distal to the lens, composed of an outermost layer of ectoderm and an ingrowing layer of mesodermal cells.

Sclera and *choroid:* not yet formed out of mesenchyme surrounding the optic cup.

Nasal organ:

Nasal pit: blind, thick-walled ectodermal sac ventrolateral to the telencephalic hemisphere. Trace.

External naris: opening of the nasal pit to the outside.

Median nasal process: the bulge of mesenchyme medial to the naris.

Lateral nasal process: the bulge of mesenchyme lateral to the naris.

Cranial nerves: paired structures described as single; terminal distribution to be studied later in the exercise.

Spinal accessory (XI):

Spinal accessory nerve: Return to a frontal section through the hindbrain which shows a prominent long fibrous strand lateral and parallel to the brain; this is the eleventh nerve. In a few sections caudal in the series the nerve "breaks" into two parts, a dorsal part which can be traced for a long distance as a fibrous body lying adjacent to the dorsolateral surface of the brain and anterior part of the spinal cord. The ventral segment will be traced through the mesenchyme later.

Accessory ganglia and *Froriep's ganglion:* In sections passing longitudinally through the spinal accessory nerve, noted above, observe an elongated ganglionic mass, the accessory ganglia, lying lateral to the eleventh nerve. This body becomes more prominent in posterior sections and is here known as Froriep's ganglion. These ganglia are considered the sensory parts of the spinal accessory and hypoglossal nerves; the ganglia are not present in the adult as such (i.e., as discrete bodies) although the eleventh and twelfth nerves contain sensory components.

Vagus (X):

Jugular ganglion: Return to more dorsal sections again

181

to find a large ganglionic mass lateral to the eleventh neuromere of the brain and anterior to the spinal accessory; this is the jugular ganglion of the vagus. Trace it ventrally and note that it gives place to a pale body of nerve fibers, the tenth nerve.

Nodose ganglion: Continue to trace the tenth nerve carefully. It is the middle member in a string of three bodies, the first being the spinal accessory nerve and the third body, the glossopharyngeal nerve (to be described). All lie along the inner border of the large anterior cardinal vein (see fig. 60). Soon the tenth and eleventh nerves lie side by side; then the eleventh nerve "bulges" into the anterior cardinal vein and later "moves" laterally. At about this point the vagus now becomes a large dark ganglionic mass. This is the nodose ganglion, which contains the nerve cell bodies of visceral sensory neurones whereas the jugular ganglion contains the nerve cell bodies of somatic sensory neurones.

Glossopharyngeal (IX):

Superior ganglion: Return to dorsal sections again to find a ganglion, smaller than the jugular ganglion, lying between the latter and the auditory vesicle. This is the superior ganglion of the ninth nerve which contains nerve cell bodies of somatic sensory neurones.

Petrosal ganglion: Trace the ninth nerve, as with the vagus, until a second dark ganglionic mass appears not far from the nodose ganglion. This is the petrosal of the ninth which contains nerve cell bodies of visceral sensory neurones.

Acoustico-facialis ganglion (VII and VIII): a very dark body anterior to the auditory vesicle. Trace ventrally to note that it becomes separated into two distinct masses.

Auditory ganglion (VIII): the part of the acoustico-facialis complex adjacent to the auditory vesicle.

182

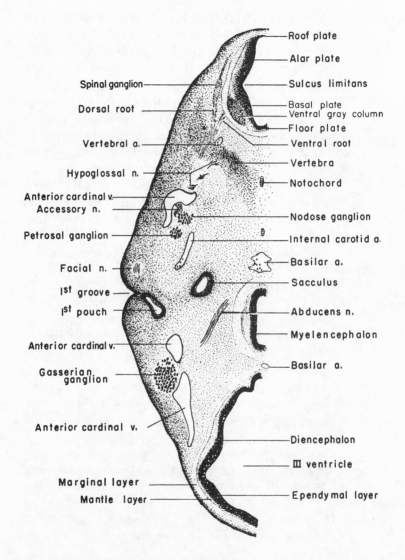

Fig. 60. Cross section of 10 mm. pig embryo at level of first pouch.

183

Geniculate ganglion (VII): the anterior part of the acoustico-facialis complex. Note the bundles of nerve fibers emerging from the hindbrain to form the motor elements of the facial nerve.

Trigeminal (V): Again in dorsal sections note the *semilunar* (or *gasserian*) *ganglion* of the fifth cranial nerve at the widest part of the hindbrain (posterior border of neuromere 7). Trace the ganglion ventrally. Note bundles of motor fibers emerging from the floor of the metencephalon.

Hypoglossal (XII): This largely motor nerve is represented by numerous twigs emerging from the ventrolateral wall of the myelencephalon in sections about to pass through the floor of the hindbrain.

Abducens (VI): In sections at the level of the sacculus a small motor nerve will be seen running through the mesenchyme between the ear and the brain. This is the sixth cranial nerve. Note its connection to the hindbrain.

Trochlear (IV):* a very small motor nerve, poorly developed at this stage, emerging from the roof of the mesencephalon at its posterior boundary and passing ventrally in the mesenchyme adjacent to the isthmus of the brain.

Oculomotor (III): Search sections showing the mesencephalon unconnected to the hindbrain for two small bodies attached to the floor of the mesencephalon. These are the third cranial nerves.

Optic (II): represented at this stage of development by fine threads (axones) emerging from the inner surface of the retina and converging into the choroid fissure, along which they will run to the brain.

Olfactory (I):* faint bridge between the medial wall of the nasal pit and the anteroventral wall of the telencephalon.

Digestive and respiratory systems: Structures will be described as though the slides are oriented on the stage of the microscope with sections of the spinal cord appearing upper-most in the field. Paired structures will be described as single.

First pharyngeal (hyomandibular) pouch and *furrow:* In sections passing through the ventralmost part of the hind-brain and auditory vesicle, note the first pharyngeal pouch, lined with thick entoderm, and the furrow in the thin ectoderm. Is a closing plate present? The mesenchyme adjacent to the anterior border of the pouch and furrow is the *first visceral* or *mandibular arch.*

Second pharyngeal pouch and *furrow:* Proceed caudally in the series. There appears now a second pharyngeal unit similar to but behind the first unit, that is, nearer the end of the section containing the spinal cord. Trace posteriorly in the series to observe the connection of the first two pouches to each other. Is there a closing plate between the second pharyngeal pouch and furrow? The mesen-chyme between the first and second pouches is the *second visceral* or *hyoid arch.*

Pharynx: the large median cavity into which the pharyn-geal pouches open. In tracing posteriorly in the series (ventrally in the pharynx) the floor of the pharynx "di-vides" the pharynx into two parts, the anterior one which may be traced into the stomodeum and the posterior one which will be traced later into the trachea and oesophagus. This picture of divided pharynx is owing to the bending of the pharynx as a result of flexure of the head. Confirm this point by applying a straightedge to a text figure (e.g., Arey's Developmental Anatomy, fig. 601 or fig. 98 of Patten's Embryology of the Pig).

Stomodeum: the median chamber into which leads the anterior part of the pharynx just described. Proceeding further in the series (ventrally in this region of the em-bryo), the stomodeum or oral aperture becomes continu-ous to the outside laterally by a narrow cleftlike passage. Now the anteriormost part of the head, which is greatly

185

bent by cranial flexure, is completely separated from the rest of the body.

Rathke's pouch: a flattened ectodermal vesicle, at the base of the diencephalon, which opens into the stomodeum.

Maxillary process: the lateral bulge above the eye and below the cleftlike communication of the stomodeum to the outside, the slide being oriented with the spinal cord away from the observer.

Mandibular process: the bulge above the maxillary process and the lateral connection of the stomodeum to the outside. Both maxillary and mandibular processes are parts of the first or mandibular arch which is bent into an inverted U (see fig. 598 in Arey or fig. 32 in Patten's Embryology of the Pig).

Tongue: Return to sections in which the pharynx first appears and shows connections to the first and second pouches. Bear in mind that sections through the anterior part of the pharynx are cut in a frontal plane. Consult figure 98 of Patten's Embryology of the Pig. Note an "island" in the pharynx. This is part of the tongue. Trace it ventrally (i.e., posteriorly in the series) to observe that it fuses with the floor of the second and third visceral arches. The following rudiments of the tongue may be distinguished.

Tuberculum impar: the pointed tip of the "island" mentioned above. In more ventral sections where the island becomes fused with the second and third visceral arches the tuberculum impar is represented by the small median process seen in figure 61. It is regarded as part of the first visceral arch.

Copula: the future root of the tongue, represented by the medial part of the second visceral arch.

Lateral tongue swellings: Trace the tongue ventrally noting that the tuberculum impar, just before it "dis-

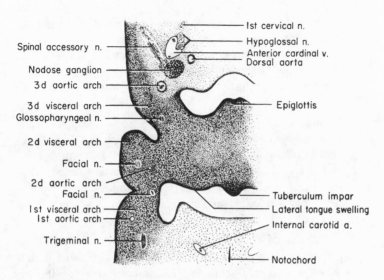

Fig. 61. Frontal section (cross section) through pharynx of 10 mm. pig embryo.

appears," is connected on each side with a bulge on the mandibular arch. These are the lateral tongue swellings which will later fuse in the midline anterior to the tuberculum impar and form the body of the tongue. The tuberculum impar will thus lie between the body and the root (copula) of the tongue. Consult figure 602 of Arey, or figure 173 of Patten's Embryology of the Pig, or figure 251 of Patten's Human Embryology.

Epiglottis: medial swelling from the posterior border of the third visceral arch (see fig. 61). It is considered a derivative of the third and fourth visceral arches.

Fourth pharyngeal furrow: a recess from the posterior border of the third pharyngeal furrow. The fusion of the third and fourth furrows into a common opening, known as the *cervical sinus* (see Arey, fig. 598), is probably owing to flexure of the embryo.

187

Thyroid: dark, irregular body in the mesenchymatous floor of the pharynx at the level of the hyoid arch. A small dorsal stub of thyroid represents the remnant of the thyroglossal duct. Its connection with the pharynx, however, has been lost. Under high power a faint suggestion of follicles may be seen.

Parathyroid of third pouch (G. *para* = prefix meaning alongside of): In sections passing through the dorsal part of the third pouch may be seen a light thickening on the anterior wall of the pouch, adjacent to a large blood vessel (third aortic arch). This is the primordium of the inferior parathyroid. Consult figure 613 of Arey or figures 96–98 of Patten's Embryology of the Pig.

Thymus of third pouch: usually not differentiated as yet; represented by the ventral wing of the third pouch which in ventral sections (i.e., more posterior in the series) appears as an isolated elliptical vesicle. This vesicle can be traced for several sections; note that it extends ventrally and medially.

Arytenoid swellings (G. *arytainoeides* = shaped like a ladle): In sections passing through the ventral part of the third pouch (thymus) the pharynx has a narrow U-shaped cavity. This appearance of the pharyngeal lumen is owing to two prominent folds rising from the pharyngeal floor. These are the arytenoid swellings, derivatives of the fourth and fifth visceral arches, which will later form cartilaginous supports for the larynx.

Glottis: the V-shaped opening between the arytenoid folds. It communicates with the *larynx,* an ill-defined region of the respiratory system represented by the slightly dilated ventral end of the laryngotracheal groove between the arytenoid folds. The cavity of the larynx is obliterated at this stage by the apposition of its walls.

Fourth pharyngeal pouch: a small lateral recess from the U-shaped pharyngeal cavity. (See Arey, fig. 615 insert or fig. 69 of Patten's Embryology of the Pig.)

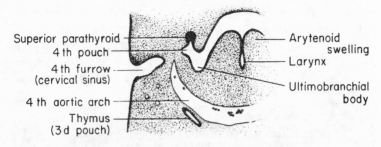

Superior parathyroid
4th pouch
4th furrow
(cervical sinus)
4th aortic arch
Thymus
(3d pouch)

Arytenoid
swelling
Larynx
Ultimobranchial
body

Fig. 62. Cross section through the fourth pharyngeal pouch of 10 mm. pig embryo.

Parathyroid of fourth pouch: The superior parathyroid is just beginning to differentiate; it is represented by the epithelium of the dorsal part of the fourth pouch, which sometimes forms a very small diverticulum (see fig. 62).

Ultimobranchial body (L. *ultimus* = last + G. *branchia* =gills) or *lateral thyroid:* small outpocketing from the ventral wall of the fourth pouch. In a few sections caudally in the series it appears as an isolated circle of epithelium. It is thought to represent the fifth pharyngeal pouch (see fig. 62).

Thymus of fourth pouch: usually not yet differentiated. If present it is represented by the small part of the fourth pouch extending lateral to the superior parathyroid.

Review the derivatives of the pharynx; see p. 158.

Esophagus: flattened thick-walled part of the foregut posterior to the pharynx; in more posterior sections the esophagus becomes circular in cross section.

Lower jaw: Fusion of the mandibular processes to form the lower jaw may be seen to advantage in sections passing through the esophagus. Lateral tongue swellings may still be seen on the dorsal surface of the jaw.

Upper jaw: Above the lower jaw on each side may be seen the primordium of the upper jaw formed by the fusion of the maxillary process and the median nasal process.

189

Trachea: thick-walled entodermal tube ventral to the esophagus.

Bronchial buds or *lung buds:* outgrowths from the trachea. The stems of the buds, here very short, become the bronchi whereas the expanded ends of the buds become the lobes of the lung.

Apical (eparterial) bronchial bud: the anteriormost bud appearing as a small evagination from the right side of the trachea. This bud will form the eparterial bronchus (so named because it will later lie upon the pulmonary artery) and the upper lobe of the right lung.

Primary bronchial buds: In more posterior sections these buds appear as evaginations from the caudal end of the trachea. The right primary bud shows a further bifurcation into anterior and posterior buds which will form respectively: middle bronchus and lobe of the lung, and lower bronchus and lobe of the lung. The left primary bud is likewise divided into two buds, anterior and posterior, which will give rise to upper and lower bronchi and lobes of the lung (see fig. 63).

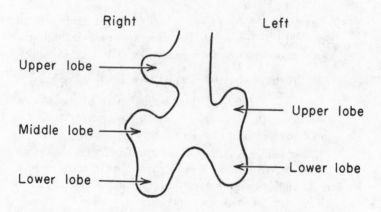

Fig. 63. Lung buds in the 10 mm. pig embryo, ventral view.

190

Stomach: Trace the esophagus until the entodermal tube begins to enlarge and elongate; this is now stomach.

Liver: the large mass ventral to the stomach consisting of paired dorsal and ventral lobes.

Liver cords: the strands of liver cells.

Sinusoids: the numerous small circulatory passages between the liver cords.

Duodenum: Trace the elongated stomach until it gives place to the oval-shaped duodenum.

Bile duct or *ductus choledochus* (G. *chole* = bile + *dochos* = containing): a large thick-walled canal first seen embedded in the liver substance ventral and to the right of the stomach. Trace it posteriorly to its entrance into the right ventrolateral wall of the duodenum (see fig. 64).

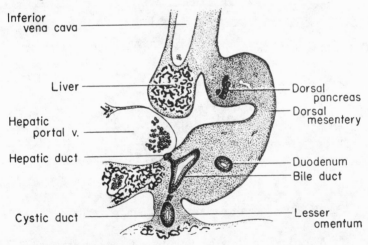

Fig. 64. Formation of bile duct in 10 mm. pig embryo.

Hepatic ducts: numerous small channels from the liver opening into the bile duct.

191

Cystic duct: Before the bile duct is seen to enter into the duodenum an elliptical or rounded tubule is given off ventrally. This is the cystic duct.

Gall bladder: Trace the cystic duct until it expands into an elongated vesicle, the gall bladder. Examine under high power.

Pancreas (see fig. 65):

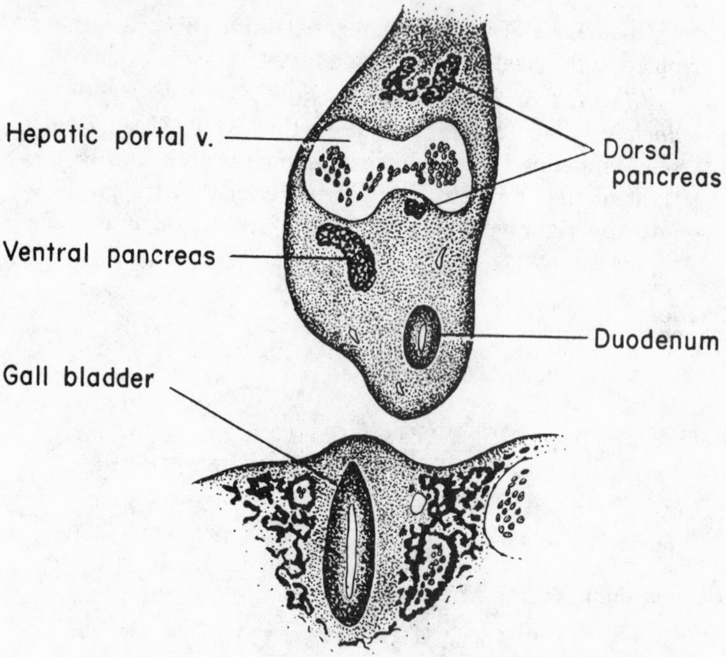

Fig. 65. Formation of pancreas in 10 mm. pig embryo.

Dorsal pancreas: large dark mass in the dorsal mesentery (*mesoduodenum*) above the duodenum. Examine under high power. Note that alveoli are beginning to form, interconnected by ducts. Trace the dorsal pancreas posteriorly to observe it open into the dorsal wall of the duodenum by its *duct of Santorini* (after Giovanni Santorini, Italian anatomist, 1681–1737).

Ventral pancreas: Return to sections passing through the entrance of the bile duct into the duodenum. Proceed posteriorly again noting that as the bile duct disappears a dark body persists a short distance to the right of the duodenum. This is the duct of the ventral pancreas, known as the *duct of Wirsung* (after Johann Georg Wirsung, Bavarian anatomist, d. 1643). The lumen may be seen under high power. Trace the duct in posterior sections. It "moves" farther to the right and dorsally and leads to the ventral pancreas, similar in structure to the dorsal pancreas. Trace farther to note the close relationship between dorsal and ventral pancreas.

(The posterior parts of the digestive system will be studied later.)

Circulatory system: paired structures described as single.

Anterior cardinal vein: In sections passing through the sacculus of the auditory vesicle identify this large vein lateral to the otocyst and acoustico-facialis ganglion. Observe that in a few sections posteriorly it seems to be "divided" by the seventh nerve. The anterior part continues forward in the head of the embryo medial to the semilunar ganglion and dorsally along the side of the diencephalon; observe that it receives medial and lateral tributaries. The posterior division courses back in the embryo lateral to the posterior cranial nerves and ganglia.

Internal carotid artery: A longitudinal section through this vessel, smaller than the cardinal vein, may be usually found just medial to the petrosal ganglion. In the next few posterior sections it becomes divided. The anterior part may be traced forward into the head of the embryo. Note that it may receive a small vessel from the second

193

visceral arch, the remnant of the *second aortic arch*. Trace farther to observe the internal carotid extend dorsally past Rathke's pouch and along the side of the diencephalon, then forward in series alongside the *infundibulum* to join its mate under the mesencephalon. Return now to the longitudinal sections through the internal carotid near the petrosal ganglion. Proceed to trace caudally the posterior part of the vessel. It receives the third aortic arch.

Paired dorsal aorta: Continuing caudally from the entrance of the third aortic arch just noted the vessel is no longer internal carotid but one of the paired dorsal aortae. Trace the dorsal aorta farther; it lies medial to the nodose ganglion and the anterior cardinal vein.

Third aortic arch: Leaving the dorsal aorta for the moment trace the large third aortic arch down through the substance of the third visceral arch. The two third aortic arches fuse together just posterior to the thyroid to form the *ventral aorta*. Actually, of course, the arches arise from the ventral aorta.

External carotid artery: Just before the third aortic arches fuse midventrally note that a small vessel, the external carotid artery, arises from the anterior wall of the arch and extends forward into the lower jaw or mandibular arch.

Fourth aortic arch: Considering the dorsal aorta again observe that it becomes vertically elongated and that the ventralmost part is "pinched off" as the fourth aortic arch. Trace the arch ventrally to its junction with its mate and the ventral aorta. Sometimes a single section will show the entire fourth arches connecting dorsal and ventral aortae, thus forming a large U.

Systemic trunk of the conus: Trace the ventral aorta posteriorly. On the right side there persists a large thick-walled vessel which soon becomes surrounded by pericardial cavity. This is the systemic or aortic trunk of the conus of the heart.

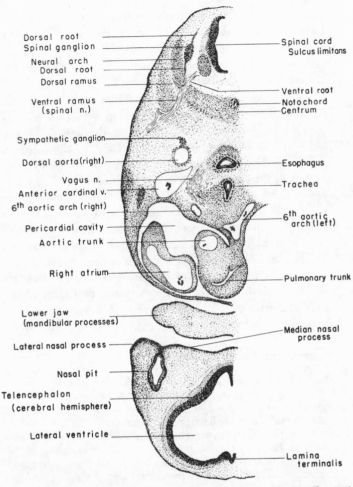

Fig. 66. Cross section of 10 mm. pig embryo at level of sixth aortic arches.

Sixth aortic arch: Return to the section showing the connection of the fourth aortic arch to the dorsal aorta. With attention on the dorsal aorta proceed caudally in the series. Note that another large vessel is received. This is the sixth aortic arch. Trace it ventrally to observe that it joins its mate. Is there a difference in size of the two sixth aortic arches?

Fifth aortic arch: if present, a small channel connected at each end of the sixth aortic arch.

Pulmonary trunk of the conus: The sixth aortic arches just traced form the arms of a Y-shaped figure, the stem of which emerges from a large vessel to the left of the systemic trunk. This is the pulmonary trunk of the conus of the heart. Tracing farther note that owing to the spiraled bulbar septum the pulmonary trunk lies ventral, the systemic trunk dorsal in position. The two trunks may be connected still by a narrow channel or in some specimens (younger) the two trunks may not be separated at all. The *semilunar valves* in the systemic and pulmonary trunks will later differentiate from the walls of these great vessels.

Conus: This chamber of the heart consists of the two trunks, described above. It lies between two large chambers, the atria of the heart, which will be ignored for the moment. Continue tracing the twisted conus posteriorly.

SPIRALING IN THE CONUS

As observed here (and later to be seen in the dissection of the fetal pig) the pulmonary and aortic trunks appear intertwined through 180°, the former being twisted ventrally and to the left, the latter dorsally and to the right. This torsion is not the result of an actual rotation of the conus (bulbus) either before or after subdivision into the two trunks, but it is due to a spiraling of the bulbar septum internally. What causes the bulbar septum to spiral?

A recent investigation of human cardiogenesis by Pieter de Vries demonstrated that the spiraling is probably owing to the hemodynamic action of the streams of blood ejected from the ventricles at each systole. He studied with models the configurations of a fluid resulting from the intersection of two streams. He found that a left stream joining slightly behind a right stream causes a clockwise spiral. His analysis of the developing human heart showed that the morphological positioning of the ventricles would give a junction of the left and right ventricular streams in such a manner as to produce a clockwise spiral of the blood in the conus. Reverse positioning of the joining streams in the model causes a counter-clockwise spiral.

The torsion of the bulbar septum is an excellent example of the importance of fluid dynamics in the morphogenesis of the circulatory system.

Right ventricle: The pulmonary trunk is carried to the right and is connected to the right ventricle, a thick-walled chamber of the heart.

Left ventricle: The systemic part of the conus, now twisted to the left, communicates with the left ventricle. Study the structure of one of the ventricles.

Trabeculae carnae: loosely arranged bundles of muscle cells in the ventricular wall. Examine under high power.

Endocardium and *epicardium:* inner and outer epithelial coverings of heart muscle or *myocardium.* Examine under high power.

Interventricular septum: the pillar of tissue incompletely separating right and left ventricles.

Interventricular foramen: the connection between the two ventricles.

Atrio-ventricular canals:

Right canal: a narrow passage connecting the right atrium and the right ventricle.

Left canal: a similar passage between left atrium, a small part of which may be seen at posterior levels, and left ventricle.

Endocardial cushions: the thick processes, guarding the atrio-ventricular canals, which are beginning to form the auriculo-ventricular valves.

Tricuspid valves: represented here by the endocardial cushions about the right canal.

Bicuspid or *mitral valves* (G. *mitra* = belt or turban): represented by the endocardial cushions about the left canal.

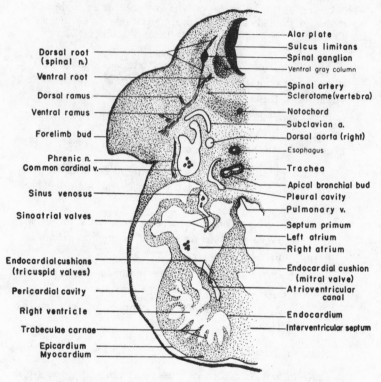

Fig. 67. Cross section of 10 mm. pig embryo at level of the heart.

Right atrium: the large chamber of the heart dorsal to the right ventricle from which leads the right atrio-ventricular canal described above.

Sino-atrial valves: two thin flaplike structures extending into the atrium from its dorsal wall. These valves guard the entrance of the sinus venosus into the right atrium.

Interatrial septum or *septum primum* (L. *septum* = wall; *primus* = first): the partition between the two atria.

Foramen ovale (L. *foramen* = hole; *ovale* from *ovum*, meaning egg-shaped): opening in the interatrial septum permitting communication between the two atria. To observe the foramen it may be necessary to move forward in the series a number of sections.

FUNCTIONAL SIGNIFICANCE OF FORAMEN OVALE

Approximately 75 percent of the blood from the inferior vena cava and about 25 percent of that from the superior venae cavae flows through the foramen ovale (here used as the passageway from right to left atria). This assumption is based upon the studies of Newton Everett on the dog fetus. As the former stream bears the freshly oxygenated, nutritionally enriched, and purified blood from the placenta it is significant that most of it passes to the left side of the heart which, of course, feeds the arteries to the head. This arrangement assures channeling of the best blood to the vital centers of the fetus. Second, the shunt of blood through the interatrial partition provides adequate exercise for the left side of the heart. Without the moulding hemodynamics of a good flow of blood through left atrium and ventricle these chambers would not develop properly and consequently be unprepared to handle the sudden additional load placed upon them at the time of birth. On the other hand, the right atrium and ventricle would, without the foramen ovale, be overdeveloped. See p. 237 for a discussion of changes in the circulatory system at birth.

Sinus venosus: the small chamber immediately dorsal to the sino-atrial passage.

Common cardinal or *duct of Cuvier:* On the right side, the large vessel dorsal to and continuous with the sinus venosus. Identify the left common cardinal in a similar position on the other side of the embryo. Both lie lateral to the dorsal aortae. Incidentally compare the walls of the aortae with those of the common cardinals. Is there a difference in size between the two aortae? To establish a connection between left common cardinal and sinus venosus it will be necessary to trace carefully the ventral part of this great vein down into the pericardial cavity and behind (posterior to) the left atrium. Sections in which union of the left common cardinal and sinus venosus is made lie toward the posterior end of the heart. Usually only a small piece of right atrium may be seen here, nothing of the left atrium, but of course still large sections of the ventricles.

Pulmonary veins: Return to sections showing a small left atrium with the left common cardinal lying next to its lateral wall. A narrow "diverticulum" from the left atrium may be seen extending dorsally and just to the left of the interatrial septum. This is the pulmonary vein. Trace it dorsally to observe the bifurcation of the vein into the right and left pulmonary veins. Trace these vessels as far as possible.

Pulmonary arteries:* Return now to sections showing the sixth aortic arches. From the inner side about midway on each arch is given off a small vessel, the pulmonary artery, which can be traced in favorable specimens posteriorly toward the lung buds in the mesenchyme below the trachea.

Postcaval vein (posterior vena cava): Locate again the sections showing entrance of the left common cardinal vein into the sinus venosus. Actually this is not the sinus venosus but the anteriormost part of the postcaval vein. Trace the postcaval forward to the sinus venosus identified earlier as the small chamber immediately dorsal to the sino-atrial passage. Now trace the postcaval posteriorly. It becomes a large oval channel, embedded in a septum of mesenchyme extending transversely across the body. This partition is the *septum transversum* to be studied later. In a few sections more posteriorly the postcaval is surrounded by liver.

Ductus venosus: A large channel in the liver may soon be seen emptying into the postcaval from the left. This is the ductus venosus. Note the relationship of the liver sinusoids to the ductus venosus and to the postcaval. In more posterior sections the ductus venosus is separated from the postcaval. The latter now lies to the right of the stomach partly embedded in liver and supported by the *caval mesentery;* the former lies ventral to the stomach and entirely within the liver. The use of the term ductus venosus for this hepatic channel carrying primarily umbilical blood differs from the usage of the term in the chick (p. 162).

Hepatic portal vein: Examine sections passing through the region of the gall bladder. Observe a large channel in the small upper right lobe of the liver. This is the hepatic portal vein. The postcaval lies far dorsally, above the dorsal pancreas; the ductus venosus now lies in the left lower lobe of the liver. Trace the hepatic portal forward to show its relation to the postcaval and posteriorly a short distance to see its association with pancreas and duodenum. (The posterior parts of the circulatory system will be studied later.)

Urinary system: Review vertebrate nephric system (p. 142).

Mesonephros: large bodies of coiled tubules and blood vessels, one on each side of the dorsal aorta, projecting into the body cavity. The organs are first seen in sections passing through the lungs. In proceeding caudally the mesonephros becomes exceedingly large.

Mesonephric tubules: the numerous canals, cut at various angles, making up the bulk of the mesonephros. Examine the epithelial lining under high power.

Bowman's capsules: large thin-walled vesicles, medially situated in the mesonephric bodies. Find the entrance of a capsule into a mesonephric tubule.

Glomeruli (L. diminutive of *glomus* = ball): the large capillary masses extending into the Bowman's capsules.

FUNCTION OF FETAL NEPHROI

One might question the functional activity of the mesonephros and also that of the metanephros before birth in the fetus of a placental mammal, considering the fact that the placenta provides for the elimination of nitrogenous and other wastes via the maternal blood stream. There is evidence, however, that the fetal nephroi do function, although to varying degree in different mammals, as shown by the following observations and experiments.

1. Urea can be demonstrated in the amniotic and allantoic fluids (as early as two and a half months in the human amniotic fluid). Urea would reach the former by way of the urethra, the latter through the urachus (allantoic stalk).

2. Tracers (e.g., dyes or radioactive substances) injected into the maternal circulatory system may be later identified in the allantoic and amniotic fluids.

3. Occlusion of the nephric duct leads to dilatation of the nephros anterior to the obstruction. Similarly, blockage of the fetal ureter is followed by hydronephrosis (G. *hydor* = water + *nephros* = kidney) of the kidney.

4. Histochemical studies demonstrate that glomerular filtration and tubular secretion are both taking place in the fetal nephroi. Localized precipitates of Prussian blue (excreted by the glomeruli) and phenol red (excreted by the tubules) are particularly convincing.

5. Fragments of embryonic nephros in tissue culture show excretory activity.

There appears to be some correlation between the degree of development of the mesonephros and the presumed efficiency of the placenta. In the pig fetus, as observed here, the mesonephros is large and active, and the allantois forms a spacious bladder. The placenta of the pig (and some other ungulates) is *epithelio-chorial* in type, the chorion of the fetus being in close contact with the uterine epithelium. In this instance the placental barrier is of maximal thickness but shows a relatively slow functional performance. The rat, on the other hand, has a *hemo-chorial* placenta (chorionic villi bathed with maternal blood) in which the barrier is of minimal thickness and of high efficiency, but its mesonephros is rudimentary and apparently functionless. The placenta of the rat and other rodents was once thought to be hemo-endothelial, that is, possessing a barrier consisting of only one layer of cells, the endothelium of the fetal capillaries. Electron microscopy has shown, however, that there is a thin chorion covering the fetal blood vessels. Thus the rodent placenta is hemo-chorial, as is that of man and most other primates. Incidentally, carnivores have an *endothelial-chorial* placenta, so-called because the chorion is in contact with the endothelium of the maternal blood vessels.

Relationship to circulatory system:

Mesonephric arteries: small vessels arising from the ventro-lateral walls of the dorsal aorta. Trace them to one or more of the glomeruli.

Posterior cardinal veins: In the anteriormost sections passing through the mesonephros the posterior cardinal veins may be readily identified. Trace them posteriorly, especially the left one. Note the drainage from the mesonephros. Soon the posterior cardinal vein becomes "divided" by a mass of mesonephric tubules. The dorsal part only is, in posterior levels, designated the posterior cardinal vein.

Subcardinal veins: The postcaval vein after "moving up" the caval mesentery to a position beneath the dorsal aorta comes into relation with the mesonephros. In this region it is actually the right subcardinal vein. Note that the vessel drains the medial parts of the mesonephros. Farther posteriorly the left (smaller) subcardinal vein may be observed. The two subcardinals anastomose freely with one another and with the posterior cardinals.

Supracardinal veins: At more caudal levels may be seen a foreshadowing of the formation of venous channels on the medio-dorsal surface of the mesonephric bodies; these vessels will be the supracardinals which anastomose with the other systems. Later the right supracardinal becomes a segment of the post-caval.

Mesonephric or *Wolffian duct:* the large oval canal at the ventral edge of the mesonephros which receives the mesonephric tubules at intervals. Move rapidly, row by row, to slides near the end of the series to find sections showing the mesonephric ducts cut longitudinally, that is, extending back to a pelvic position medial to the hindlimb buds. Use a straightedge on figure 609 of Arey at level 628 to visualize better the picture of the mesonephric duct at this level. To trace the mesonephric ducts farther proceed from this point *forward* in the series. The ducts appear pear-shaped or oval in section and are lined with relatively thin epithelium.

Urogenital sinus (see fig. 68): Trace the mesonephric ducts farther by moving anteriorly in the series. They become connected to a narrow U-shaped chamber, the urogenital sinus, which represents a subdivision of the cloaca, to be examined later. Compare with figure 609 of Arey, level 626.

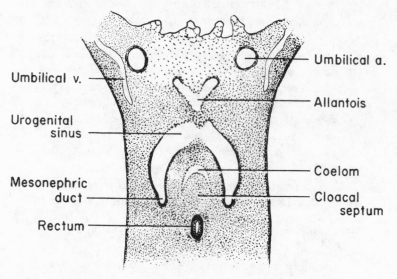

Fig. 68. Urogenital sinus in 10 mm. pig embryo.

Ureter: Return now to the section noted above in which the mesonephric ducts are cut longitudinally. Note that from the posterior ends of the ducts there arise paired thick-walled tubules, the ureters. Trace them by moving caudally in the series.

Metanephros:

Pelvis of the kidney: Trace the ureters until they expand into ovals, the future pelves of the kidneys.

Metanephrogenic mesoderm: the dark mass of mesoderm (nephrotome) about each pelvic rudiment, the forerunner of the cortex of the kidney.

Posterior part of the digestive system:

Intestinal loop: The tracing of the intestinal loop requires patience.

Cranial limb: Begin with the duodenum at the level of the pancreas. Trace posteriorly section by section; the circle of entoderm persists row after row. Soon it will begin to "move" ventrally along a median cord of mesenchyme, the *dorsal mesentery,* into a loose mesenchymatous mass, the substance of the umbilical cord. The extension of the gut into the umbilical cord is known as the *umbilical hernia* (see fig. 69). A point will soon be reached at which it will be necessary to reverse the direction of tracing in order to follow the cranial limb to the very tip of the intestinal loop where it turns back as the caudal limb. In many specimens the umbilical cord has been cut close to the body of the embryo and this union of cranial and caudal limbs may not be observed.

Caudal limb: From the tip of the intestinal loop proceed posteriorly again in the series following the caudal limb; it comes to lie after several rows of sections at the end of the ribbonlike dorsal mesentery. More caudally

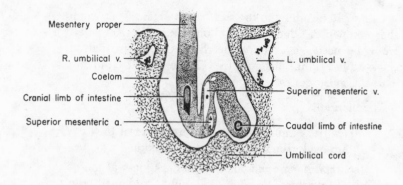

Mesentery proper

R. umbilical v.

Coelom

Cranial limb of intestine

Superior mesenteric a.

L. umbilical v.

Superior mesenteric v.

Caudal limb of intestine

Umbilical cord

Fig. 69. Umbilical hernia in 10 mm. pig embryo.

the gut becomes embedded in a thicker mesentery and is cut longitudinally. To follow the posterior part of the gut farther it will be necessary to proceed forward in the series. Trace the gut into the pelvic region of the embryo. Observe that it communicates with the urogenital sinus.

Cloaca: large common terminal chamber for digestive and urogenital systems.

Urogenital sinus: the ventral part of the cloaca, identified earlier as a narrow U-shaped chamber (see fig. 68).

Cloacal septum: a partition of mesoderm, covered with entoderm, which is dividing the cloaca frontally into the ventral urogenital sinus and the dorsal rectum (see Arey, figs. 275, 625, and 626). The septum is the peninsula of mesenchyme projecting into the urogenital sinus, making it U-shaped. In a few sections farther forward in the series the septum may appear as an "island" in the cloaca; then it disappears as its tip is reached.

Rectum: the terminal segment of the digestive canal, seen as a circle of entoderm separated from the urogenital sinus by a wedge of mesoderm, the cloacal septum. The cloacal septum is not as yet complete so rectum and urogenital sinus are confluent posteriorly.

Cloacal membrane: Continue tracing the cloaca *forward* in the series. It becomes laterally compressed and dumbbell-shaped. The ventral part is in relationship with the urogenital sinus, the dorsal part with the rectum. Soon the lateral walls of the cloaca become fused with each other and with the ectoderm; this plate of entoderm and ectoderm is the cloacal membrane. Because the plane of section is parallel to the ectoderm the relationship of the membrane to ectoderm can be seen only at the ends of the plate, until in the last few sections it clearly appears to be continuous laterally with the outer germ layer.

Tail gut: Just before the cloacal membrane is reached in the tracing above observe that a diminutive tube arises from the dorsal side of the cloaca. This is the tail gut which extends for a number of sections into the tail of the embryo.

Tail: At this point examine the structures in the tail; identify in addition to the tail gut: spinal cord, notochord, dorsal aorta (*caudal artery*), somites, and paired posterior cardinal veins.

Allantois: Return to the U-shaped urogenital sinus. Examine the sections showing a flattened canal connected to the ventral wall of the urogenital sinus. Trace this oval, the allantoic stalk, forward in the series. With what germ layer is it lined? It will soon come to lie between two vessels, the umbilical arteries (to be studied later), and farther forward it "moves" out into the umbilical cord.

Posterior part of the circulatory system:

Hepatic portal vein: Recall that the hepatic portal vein (right vitelline vein) was earlier traced caudally from the right upper lobe of the liver into the dorsal mesentery, then between the ventral and dorsal pancreas, and to a position dorsal and to the left of the duodenum, where it represents a segment of the left vitelline vein. From this point trace it posteriorly in association with the gut. It becomes divided; one division remains in the dorsal mesentery; the other division, representing the fused right and left vitelline veins, extends into the umbilical cord. Trace the common vitelline vein as far as the preparations will permit. It continues in the umbilical cord, later dividing into right and left vitelline veins which drain the yolk-sac of the pig embryo.

Superior mesenteric vein: Return to the part of the hepatic portal which remained in the dorsal mesentery. Trace it caudally to see it extend into the umbilical hernia, as the superior mesenteric vein, where it drains the intestinal loop (see Arey, fig. 623).

Superior mesenteric artery (vitelline artery): The arterial supply to the intestinal loop can be easily picked up in the last sections just studied. The superior mesenteric artery lies posterior to the vein (see Arey fig. 623). Trace this artery back in the series until a section is found in which the artery is cut longitudinally; follow the dorsal end *forward* in the series observing that the artery "moves up" the dorsal mesentery and eventually after two or three rows its origin from the ventral wall of the dorsal aorta may be seen.

Umbilical veins: Trace the ductus venosus posteriorly. After passing through the lower left lobe of the liver it leaves (enters) as the left umbilical vein. Trace this vein back into the umbilical cord. It soon becomes divided into left and right vessels. Which is larger? They may be traced as far as the preparations will permit, in the loose mesenchyme of the cord, known as *Wharton's jelly* (after Thomas Wharton, English anatomist, 1610–1673).

Umbilical arteries: These vessels were noted earlier in connection with the allantois. Relocate them, lying one on each side of the allantoic stalk. Trace them back into the embryo by proceeding caudally in the series. They come to lie at the base of the limb buds lateral to the mesonephric ducts and metanephros. Continue to trace them to their origin from the dorsal aorta.

Mesenteries, omenta, and partitioning of the *coelom:*

Mediastinum: Review the sections passing through the esophagus and trachea and lungs. The thick median partition of mesenchyme enclosing these organs is the mediastinum. Theoretically that part of the mediastinum dorsal to the esophagus is the *mesoesophagus.*

Pleural cavities: a pair of coelomic cavities, one on each side of the mediastinum, into which the lung buds develop.

Parietal pleura (L. *paries* = wall; G. *pleura* = rib or side): the lining of the lateral wall of the pleural cavity.

Visceral pleura: the medial epithelial lining of the pleural cavity which will cover the lungs.

Pericardial cavity: the coelomic cavity enclosing the heart.

Visceral pericardium: the epicardium of the heart.

Parietal pericardium: the outer epithelial lining of the pericardial cavity.

Pleuropericardial membranes (lateral mesocardia): the partitions, one on each side of the mediastinum, separating the pleural cavities above and the pericardial cavity below. These membranes are covered on their dorsal surface by parietal pleura and on their ventral surface with parietal pericardium. Later as the pleural cavities expand by burrowing into the lateral body wall these membranes form in a large part the definitive sac about the heart, the pericardium (see Arey fig. 258). Is the partition complete at this stage? Are the pleural and pericardial cavities connected anteriorly? posteriorly?

Transverse septum: the transverse partition in which the post-caval vein is embedded and into which the liver is growing. Owing to its oblique position (see Arey fig. 601) the dorsoventral extent of the septum transversum can not be seen in a single cross section. The septum separates the pericardial and peritoneal cavities. The former becomes increasingly smaller posteriorly; eventually it disappears.

Sources of the Diaphragm

The diaphragm, a muscular partition between thorax and abdomen in mammals only, is derived from the following.
1. The transverse septum which forms the ventral part.
2. Paired pleuroperitoneal folds or membranes which form the dorsal part.
3. Dorsal and ventral mesenteries, to which the transverse septum and pleuroperitoneal folds fuse, and which contribute a narrow midline stripe.
4. Lateral body wall additions to the young diaphragm formed from the above sources. The expansion of the lungs and liver, pressing into the body wall, is largely responsible for this supplement.
5. Ingrowth of skeletal muscle buds from lateral plate mesoderm or possibly from somitic myotome.

Innervation of the diaphragm by the phrenic nerves (cervical nerves 3-5, although somewhat variable; chiefly the fourth in man) occurs as the transverse septum "passes by" cervical segments 3-5 in its "caudal migration." The septum transversum does not actually migrate, of course. Its relationship to dorsal organs (e.g., neural tube, somites) merely shifts from an occipital level to that of the first lumbar segment as a consequence of differential growth rates. The dorsal organs growing more rapidly "'push forward" leaving ventral organs (e.g., heart, gut, transverse septum) at progressively more posterior levels. The phrenic nerves after entering the transverse septum, later to become the ventral part of the diaphragm, are consequently "towed" caudad and become very long.

Peritoneal cavity: the large coelomic cavity of the abdomen. The peritoneal cavity is continuous with the pleural cavities dorsal to the liver. Later these cavities will be separated by the pleuroperitoneal membranes. Identify *visceral* and *parietal peritoneum* (G. *peritonos* = stretched around).

Mesogaster or *greater omentum:* the region of the dorsal mesentery suspending the stomach. Note that the mesogaster is deflected to the left owing to the rotation of the stomach.

Caval mesentery: the mesentery to the right of the mesogaster, which swings the postcaval vein.

Omental bursa: the slitlike pocket of peritoneal cavity to the right of the stomach, separated from the peritoneal cavity by the caval mesentery.

Infracardiac recess: a blind narrow recess of the omental bursa which may be traced forward between the esophagus and the right lung bud.

Epiploic foramen (G. *epi* = upon + *plein* = to float) or *foramen of Winslow* (after Jacob Winslow, Dutch anatomist, 1669–1769): Posteriorly the omental bursa is open to the right side of the peritoneal cavity over the surface of the right upper lobe of the liver. This narrow exit of the bursa is called the epiploic foramen.

Lesser omentum: a persisting region of the *ventral mesentery,* which extends between the gut and the liver.

Gastro-hepatic ligament: a small bridge of mesoderm between the stomach and the liver which blocks the communication of the omental bursa with the left side of the peritoneal cavity.

Hepato-duodenal ligament: More posteriorly at the level of the duodenum the lesser omentum is designated the hepato-duodenal ligament. Note that the bile duct runs through this ligament.

Falciform ligament (L. *falx, falcis* = sickle + form): the broad attachment of the liver to the ventral body wall. This becomes considerably narrower later in development. The falciform ligament is that part of the ventral mesentery between the liver and the ventral body wall.

Mesoduodenum: the region of the dorsal mesentery suspending the duodenum. The pancreas grows into the mesoduodenum.

Mesentery proper: the region of the dorsal mesentery supporting the cranial limb of the intestinal loop (future small intestine).

Mesocolon: the region of the dorsal mesentery suspending the caudal limb of the intestinal loop (the future colon or large intestine).

Mesorectum: theoretically the mesoderm in the midline dorsal to the rectum.

Miscellaneous structures:

Spinal cord: Select sections from the cervical region for study.

Gross features: Identify roof plate, floor plate, sulcus limitans, alar and basal plates (see p. 155).

Layers of the *neural tube:* Identify the following in the spinal cord, then compare the structure of the cord with that of the brain.

Marginal layer: outer light nonnuclear layer. Examine under high power to see the fibrous nature of the layer.

Neuroglia (supporting cells) the occasional cells seen in the marginal layer.

Ependymal layer: the wide, very darkly nucleated inner layer.

Mantle layer: the middle layer, not so darkly nucleated as the ependymal layer. The enlarged ventral regions of the mantle layer, one on each side of the cord, represent the developing ventral gray (motor) columns.

External and *internal limiting membranes:* the fine membranes forming respectively the outer and inner borders of the wall of the spinal cord.

Spinal nerves: Select sections from the level of the stomach for study of one or more pair of spinal nerves.

Dorsal root ganglion: the large, darkly nucleated body (paired) lateral to the spinal cord.

Dorsal root (sensory) of a spinal nerve: the bundle of fibers which may be followed dorsally from the dorsal root ganglion into the dorsolateral wall of the spinal cord and ventrally from the ganglion to the junction with the ventral root.

Ventral root (motor) of a spinal nerve: the bundle of fibers emerging from the ventrolateral wall of the spinal cord and running down and out to join the sensory root.

Branches of a spinal nerve:

*Dorsal ramus:** Immediately distal to the junction of dorsal and ventral roots a small bundle of fibers may be seen passing dorsolaterally at an acute angle to supply the dorsal organs (skin and muscles).

*Ramus communicans:** After giving off the dorsal ramus the rest of the spinal nerve passes ventrolaterally as a large bundle of fibers for a short distance and then gives off medially the ramus communicans which passes medioventrally to the *sympathetic ganlion,* a vague, dark, granular body situated above the dorsal aorta.

Ventral ramus: the continuation of the spinal nerve ventrolaterally beyond the point of junction with the ramus communicans.

Somite: Examine the last few rows of the series in which the sections are frontal slices from the back of the embryo. Here the somites and related structures show to excellent advantage. Study first a dorsal level (toward the end of the series) and then a ventral level (forward in the series a little).

Dermatome: the outermost part of a somite seen as a gently curved band beneath the ectoderm of embryonic epidermis. Note the external grooves in the ectoderm which correspond with the intersegmental lines.

Myotome: the lighter rhomboidal mass medial to and corresponding with the dermatome. Examine the myoblasts under high power.

Sclerotome: a roughly spool-shaped mass medial to the myotome. The sclerotome is widest at its ends; in the middle it is narrow as though compressed between the belly of the myotome and the spinal ganglion or nerve. Note that the posterior half of each sclerotome is denser and darker; this region is the *caudal sclerotome.* The lighter, less dense, anterior region is the *cranial sclerotome.* Later as a result of reorientation the caudal sclerotomes of one pair of somites will fuse to the cranial halves of the next posterior pair of somites to form the primordium of the vertebra. What will then be the segmental arrangements of myotomes and vertebrae?

Spinal nerves: Ventral to the spinal ganglia note the light fibrous bodies, the spinal nerves. Observe the motor fibers emerging from the spinal cord.

Myosepta and *intersegmental arteries:* Note in more ventral sections dark streaks marking the intersegmental lines. These are in part connective tissue cells forming the myosepta and in part intersegmental arteries. Trace the intersegmental arteries to their origins from the dorsal aorta. Observe the relation of sclerotomes and myotomes to the myosepta.

214

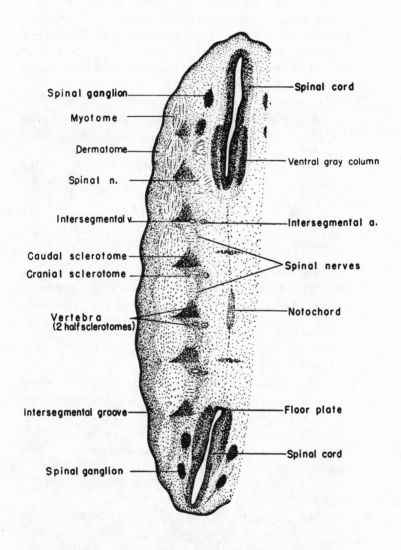

Fig. 70. Frontal section (cross section) of 10 mm. pig embryo at a dorsal level of the trunk at caudal flexure.

215

Intersegmental veins: larger thin-walled vessels, usually seen as empty circles, lying anterior to the intersegmental arteries. Trace these veins into the cardinal systems of the mesonephros.

Notochord: A good strip of notochord will appear in one of the sections in this region.

Anterior somites (seen in cross section): Study the somites at the level of the forelimb bud. Identify dermatome, myotome, and sclerotome. What is the difference in picture between a section through the middle of a somite and one at the intersegmental level? Observe the dense caudal sclerotome and the dim outline of the future vertebra. Identify rudiments of the *centrum,* the scleroblasts about the notochord, and *neural arches,* the dorsal wings arching up along the sides of the spinal cord and ganglia.

Forelimb buds: These large rudiments of the future forelimbs, seen in passing earlier, may be studied more closely now.

Ectoderm: Note that the ectodermal covering is a layer of cuboidal cells which forms a ball at the tip of the limb bud.

Mesoderm: The core of the limb bud is undifferentiated mesoderm which originated from the somatic layer of the lateral plate.

Brachial plexus: the ventral rami of several spinal nerves (certain cervical and thoracic nerves). These nerves are interconnected, hence forming a plexus. This can be determined by tracing one or more of the nerves to the forelimb.

Subclavian arteries (seventh intersegmental): Search for a section showing paired vessels from the dorsolateral wall of the dorsal aorta (or aortae) which pass up along the sympathetic ganglia and laterad in an arc above the posterior cardinal veins to enter the limb buds below the brachial plexuses. In favorable specimens the arteries can be traced a short distance in the limb buds.

Subclavian veins: Near sections showing the subclavian arteries identify the subclavian veins as large vessels dorsal to the brachial plexus which turn ventrally to open into the posterior cardinal veins. Trace the veins distally into the limb buds.

Hindlimb buds: The features of the hindlimb buds are very similar to those of the forelimb buds.

Sciatic plexus: bundles of nerve fibers passing from the spinal cord into the hindlimb buds. These are ventral rami of several spinal nerves (certain lumbar and sacral nerves).

*Iliac artery:** In favorable specimens a small artery is given off of each umbilical artery at the base of the limb bud and anterior to the sacral plexus. Usually it can not be traced far into the limb.

*Iliac vein:** small vessel draining each hindlimb bud and emptying into the posterior cardinal vein. Not always clear.

Coeliac artery: a median unpaired artery arising from the ventral wall of the dorsal aorta and passing in the mesogaster to the stomach.

Basilar artery: To trace the basilar artery is worthwhile if time permits. Pick up the basilar artery where the two internal carotids fuse under the mesencephalon (see p. 183). Trace the basilar to the hindbrain by way of a net of vessels in the mesenchyme under the isthmus. At this point reverse the direction of tracing and proceed posteriorly in the series with attention fixed on the posterior end of the basilar artery which is in relation to the hindbrain. Beneath the floor of the hindbrain the vessel becomes a large channel cut longitudinally. To trace the basilar farther one must proceed forward again in the

217

series following the posteriormost end of the artery. The vessel is obviously curved ventrally as it follows the ventral bulge of the hindbrain known as the pontine flexure. These relations are shown in figure 71. The basilar now appears a little posterior to the two sacculi of the otocysts. Continue to trace forward. The artery "moves" posteriorly toward the spinal cord. Again a somewhat irregular longitudinal section of the vessel will be encountered medial to the twigs of the hypoglossal nerves. With attention on the posterior end of the vessel, trace posteriorly. The vessel is now running under the spinal cord. It will continue under the spinal cord caudally as an ill-defined channel, the *spinal artery.*

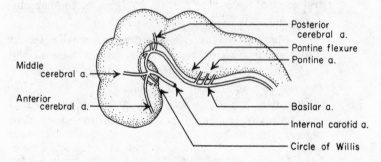

Fig. 71. Arteries to the brain of the 10 mm. pig embryo.

Vertebral arteries: This pair of vessels is not well formed yet. They originate from the subclavian arteries (seventh intersegmental arteries). Locate the sections showing the subclavian arteries. At the apex of the arches made by the subclavians may be seen in favorable specimens small vessels passing dorsally to the mesial side of the spinal nerves. These are the vertebrals. They may be traced with difficulty. They are connected farther forward with the basilar artery.

Details in the distribution of *cranial nerves:* If time

permits trace the distribution of the third, fifth, sixth, seventh, ninth, tenth, eleventh, and twelfth cranial nerves as far as possible.

Sex Differentiation

The problem of sex differentiation should not be confused with that of sex determination. The former pertains to the morphogenesis and histogenesis of the male and female reproductive system; the latter concerns the interplay of genetic factors (sex chromosomal and autosomal) which determine, in general, the direction that differentiation will take. The genetic determiners may be overridden, however, by hormonal or environmental factors.

The genital primordia are, at the outset of their development, sexless or sex indifferent. As a result of prevailing genetic, hormonal, and environmental influences either male or female differentiation will occur. (As an aside, occasionally development goes awry and sex intermediates or hermaphrodites are formed.) Each genital primordium is thus bipotential. The gonadal ridge (see table below) may differentiate into either an ovary or a testis. Emil Witschi has marshalled evidence that the developmental potentialities for femaleness reside in the gonadal cortex, those for maleness in the gonadal medulla. He postulates that there occurs an antagonism or competition between these two centers early in development. Depending upon the influences mentioned above either the cortical or the medullary organizer dominates by means of its inductive agents or embryonic hormones (presumably estrogens in the female and androgens in the male). Moreover, the prevailing organizer appears to stimulate one of the two embryonic ducts, nephric (Wolffian) or Müllerian, and the primordia of the external genitalia so that the entire genital system differentiates harmoniously. Some of the points of evidence in support of Witschi's theory of cortical and medullary factors are as follows.

1. If orchidectomy (G. *orchis* = testis + *ektomia* = excise) or castration in the male is performed on a toad a structure anterior to the testis called Bidder's organ then transforms into an ovary. The male amphibian normally possesses small non-functional oviducts. In the castrated male toad and under the stimulation of the transformed ovary these ducts become enlarged and functional, and the animal produces viable eggs. Now Bidder's organ is a *cortical* remnant of the gonadal ridge anterior to the testis, hence a point of evidence that the cortex is the seat of female factors.

2. Ovariectomy of a young domestic fowl leads to sex reversal

219

Primordium	Female derivative	Male derivative
Gonadal ridge	Ovary	Testis
1. Cortex	Germinal epithelium	Coelomic epithelium[1]
	Tunica albuginea	Tunica albuginea
2. Sex cords	Early follicles (degenerate after birth)	Seminiferous tubules
3. Medulla	Stroma	Interstitial cells
	°Rete ovarii	Rete testis
Nephric (Wolffian) duct	°Vesicular appendage	Duct of epididymis
	°Gartner's canal	Ductus deferens
		Seminal vesicle
		Ejaculatory duct
Müllerian duct		
1. Unfused part	Oviducts	°Appendix testis
2. Fused part (genital cord)	Uterus	
	Vagina[2]	°Vagina masculina
Nephric tubules		
1. Adjacent gonad	°Epoöphoron	Efferent ductules
2. Elsewhere	°Paroöphoron	°Paradidymis
		°Appendix of epididymis
External genitalia		
1. Genital tubercle	Clitoris	Penis
2. Labio-scrotal folds	Labia majora	Scrotal sac
3. Urethral folds	Labia minora	Under surface of penis
Urogenital sinus[3]		
1. Pelvic part	Vestibule (inner part)	Prostatic and membranous urethra
	Lower vaginal epithelium[2]	Prostate gland
2. Phallic part	Vestibule between labia minora	Cavernous urethra
		Bulbo-urethral glands
	Vestibular glands	Urethral glands

° Vestigial structures. Consult Arey's Developmental Anatomy or Patten's Human Embryology for their nature and position.

[1] In descending into the scrotal sac the gonad in most mammals slips along under the coelomic epithelium.

[2] Epithelium of lower vagina comes from entodermal cells migrating from urogenital sinus.

[3] Upper end of urogenital sinus plus the base of the allantoic stalk form the urinary bladder and all (female) or upper part (male) of the urethra.

in the other direction. A genetic female transforms into a male. In the female chicken and some other birds there is only one functional ovary, the left one. The right gonad is a *medullary* remnant only. Excision of the ovary in this instance apparently removes the major source of estrogen in the bird with the consequence that the right gonad, now released from inhibition, differentiates into a testis and even forms motile sperm.

3. The freemartin in cattle is an instructive experiment of nature. In instances of heterosexual twinning (a genetic male and a genetic female co-twins) with anastomosis of the two placental circulations the female twin is invariably an intersex, called a freemartin by animal husbandrymen. The gonads of the freemartin become sex reversed and likewise the system of ducts presumably as the result of androgens from the male twin circulating through the female sibling. Surprisingly, however, the external genitalia are female in character. The freemartin clearly shows the effectiveness of blood-borne principles in altering the course of sex differentiation.

4. Parabiosis (G. *para* = beside + *bios*) is an experimental procedure in which two organisms are surgically joined. Amphibian embryos are easily fused together; one can unite tailbud embryos of different sexes, species, and age. Later, the "twins" may be studied. In general, a male embryo exerts a masculinizing influence on the differentiation of the female's gonads and ducts, unless an older female embryo is joined to a younger male or the female partner belongs to a more rapidly differentiating species.

5. The administration of androgens and estrogens have been effective in sex reversing selected vertebrates from fishes to mammals. Transformation is more easily accomplished in lower vertebrates than in higher ones. For example, Witschi has produced 100 percent females in the frog *Xenopus* by treating the developing embryos with estrogen. Conversely, the treatment of embryos with androgens yields 100 percent males. For many years efforts to sex reverse a mammalian embryo by the administration of hormones to pregnant mammals met with failure. R. K. Burns has succeeded, however, in transforming genetic male opposums into females by administering estrogen to the pouched young. Bear in mind that in the marsupial birth is precocious; the new-born is still in the indifferent stage of sexual differentiation. The presumptive testis under the influence of estrogen differentiates into an ovotestis or an almost typical ovary. However, attempts to transform a presumptive ovary by the administration of androgen have yielded negative results.

B. Dissection of the Fetal Pig

1. INTRODUCTION: By now you must be weary of microscopical study. An exercise using fingers and unaided eye should be welcome. To provide some zest, as well as refreshment, to the dissection of the fetal pig, the author suggests that you imagine yourself Andreas Vesalius (1514-64) who was compelled by curiosity about the functional anatomy of the body to dare the mores of his times. He defied the threat of punishment by the Inquisition of the Church for dissecting the human body, and he questioned the authority of the ancients, such as the Greek physician Galen (130-200), whose opinions on anatomy were revered for centuries. The many years of study by Vesalius resulted in his famous book, *De fabrica corporis humani*, published in 1543, which has been characterized as "both the first great modern work of science and a foundation stone of modern biology" by Charles Singer, a noted British historian.

Nowadays, anatomy suffers no threat from an Inquisition, but it must struggle to retain respectability in the family of biological disciplines because of the disdain which some scientists, notably the molecular biologists, hold for it. Be courageous, therefore, like Vesalius, in conducting this exercise with care, with enthusiasm, and with the conviction that it may be the foundation-stone for your future studies.

2. UMBILICAL CORD: After a brief examination of the external features of the fetal pig, cut with scissors a fresh section across the umbilical cord. Note the following structures in the cord.

Wharton's jelly: the matrix of the cord.

Umbilical vein: single, large, thin-walled vessel, which may be filled with dark blood. Why is the blood no longer red? What is the fate of the vein within the fetus after birth?

Umbilical arteries: two medium-sized vessels with walls thicker than that of the umbilical vein. Explain the presence of two umbilical arteries whereas there is only one umbilical vein. What is the fate of the umbilical arteries?

Allantoic stalk: small tube situated between the two umbilical arteries at the proximal end of the cord.

Place the pig on its back and secure it to the dissecting board by passing a string tied to the right forelimb under the board to the left foreleg. In like manner anchor the hindlimbs. With scissors make a midventral incision from the pelvic region to the neck, taking care to encircle the attachment of the umbilical cord, and the external urogenital aperture of the male. Lateral incisions may now be made just posterior to the diaphragm and at the pelvis to expose the abdomen. The flaps of the body wall may be pinned back to the dissecting board or removed.

3. THE THORAX: Study first the thorax by spreading apart the rib basket. Identify the *pericardium* (pericardial sac) enclosing the heart and ventral to it the large, soft, light-colored *thymus gland;* remove them. What is the function of the thymus? How does its size in the adult compare with that in the fetus? What are the embryonic sources of the thymus? of the pericardial sac? Continue the dissection anteriorly using a probe more than forceps and scissors. Identify the *thyroid gland,* a firm, ovoid body situated in the midline and ventral to the *trachea.* Leave the thyroid in place. Identify the *larynx.* Examine the heart and identify right and left atria and the sulcus between right and left *ventricles.* Deflect the heart first to one side and then to the other to see the *lungs* clearly. Note three lobes *(apical, cardiac* or middle, and *diaphragmatic)* on both sides and an additional lobe, the *intermediate* lobe of the right lung, below the tips of the ventricles.

ARTERIES AND VEINS ANTERIOR TO DIAPHRAGM. Carefully disclose by means of probe the great veins in the vicinity of the heart, using figure 72 as a guide. The heart is shown rotated and deflected to the left and the lobes of the right lung have been pulled to the right to disclose the union of *superior* and *inferior venae cavae.* In passing note the *vagus* and *phrenic* nerves. What is the developmental origin of the phrenic

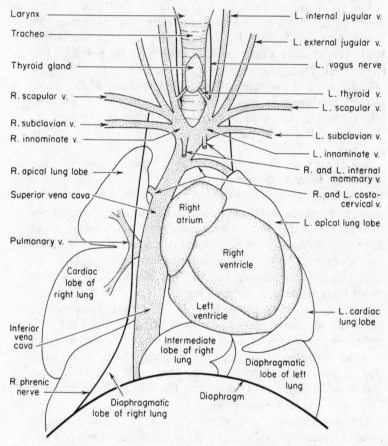

Larynx

Trachea

Thyroid gland

R. scapular v.

R. subclavian v.

R. innominate v.

R. apical lung lobe

Superior vena cava

Pulmonary v.

Cardiac lobe of right lung

Inferior vena cava

R. phrenic nerve

L. internal jugular v.

L. external jugular v.

L. vagus nerve

L. thyroid v.

L. scapular v.

L. subclavian v.

L. innominate v.

R. and L. internal mammary v.

R. and L. costo-cervical v.

L. apical lung lobe

Right atrium

Right ventricle

Left ventricle

L. cardiac lung lobe

Intermediate lobe of right lung

Diaphragmatic lobe of left lung

Diaphragm

Diaphragmatic lobe of right lung

Heart deflected to specimen's left, right lung deflected to specimen's right

Fig. 72. Venous system (in part) of fetal pig.

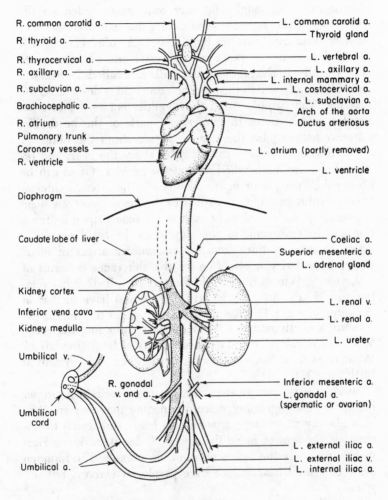

R. common carotid a.
R. thyroid a.
R. thyrocervical a.
R. axillary a.
R. subclavian a.
Brachiocephalic a.
R. atrium
Pulmonary trunk
Coronary vessels
R. ventricle

L. common carotid a.
Thyroid gland
L. vertebral a.
L. axillary a.
L. internal mammary a.
L. costocervical a.
L. subclavian a.
Arch of the aorta
Ductus arteriosus
L. atrium (partly removed)
L. ventricle

Diaphragm

Caudate lobe of liver

Coeliac a.
Superior mesenteric a.
L. adrenal gland

Kidney cortex
Inferior vena cava
Kidney medulla

L. renal v.
L. renal a.
L. ureter

Umbilical v.

R. gonadal v. and a.

Inferior mesenteric a.
L. gonadal a. (spermatic or ovarian)

Umbilical cord

Umbilical a.

L. external iliac a.
L. external iliac v.
L. internal iliac a.

Fig. 73. Arterial system (in part) of fetal pig.

nerves? What are the sources of the diaphragm? Why is the phrenic nerve so long? The only comments needed on the tributaries of the superior vena cava (also called precaval or anterior vena cava) are on terminology. A *subclavian vein* is so-called near the entrance of the vessel into an innominate vein. Distally, in the region of the axilla it is called the *axillary,* and still distally within the arm the same vein is designated the *brachial.* The *scapular* from the shoulder enters the *external jugular,* which drains the superficial parts of the head. The latter vein then joins the *internal jugular* from the brain.

Cut the superior vena cava slightly above the heart and remove the veins just studied so that the arterial system can be observed. With use of figure 73 identify the arteries anterior to the diaphragm. As in the study of the veins directions seem unnecessary. As you identify each vessel reflect upon its function and its embryonic origin. How does the *brachiocephalic* artery of the pig differ from the *innominate* artery of man? Give a speculative explanation for this difference in terms of embryology. Why is the *vertebral* artery so-called? What relationship, if any, does the blood in this vessel have to that in the internal carotid? Give careful attention to the *ductus arteriosus,* a continuation of the pulmonary trunk into the aorta. What is the role of the ductus arteriosus in fetal circulation? What is its fate? Review changes in the circulatory system at birth (p. 237).

DISSECTION OF THE HEART. To place yourself in a proper mood for this engrossing exercise, imagine that you are William Harvey dissecting a mammalian heart and trying to understand the flow of blood through it. So far as we know Harvey studied only the hearts of adult vertebrates. The foramen ovale of a fetal heart might have perplexed Harvey, but not you.

Carefully sever the great veins and arteries just beyond their connections to the heart, thereby permitting the removal of the organ from the thoracic cavity. Gently pick away the latex and blood from the orifices of the vessels. A stream of water may be useful in clearing the passageways from time to time during the dissection. In handling the isolated heart use the *pulmonary trunk* as a good point of reference for orientation (see fig. 73). Insert scissors into the pulmonary trunk and open it lengthwise along its ventral surface. Continue the incision into the right ventricle. Observe the three *right semilunar valves* at the exit of the trunk from the ventricle. One may be damaged by the incision. What is their developmental origin? How do they function? Probe the *superior* and *inferior venae cavae* to observe their entrance into the right atrium. Slit the thin wall between them. This thin wall was once the sides of what embryonic structure? Note the entrance of the *coronary sinus* into the right atrium under a ridge on the medial wall of the chamber. The sinus may be observed externally as a vessel running in the sulcus between the atria and the ventricles. Insert a probe into it. What is the developmental origin of the coronary sinus? What is its function?

Carefully remove latex from the medial side of the entrance of the inferior vena cava to reveal the *foramen ovale*. Pass a probe through it into the left atrium. What are the advantages of the foramen ovale in the fetal circulation? What would be the disadvantage of persistence of its patency in the adult? Note the thin *septum primum* that acts as a flap valve and the sturdy *septum secundum*. Picture the entrance of blood into the right atrium from the inferior vena cava with the larger part of the stream passing via the foramen ovale into the left atrium. But observe also how part of the stream would flow into the right ventricle. Now examine the projectory of blood entering the right atrium from the superior vena cava. Insert scissors through the right atrio-ventricular canal and cut along the back (dorsal) side of the heart until the right ventricle is opened. Examine the three *tricuspid valves* in the canal. One may have been damaged by the incision.

Explore the entrance of the *pulmonary vein* into the left atrium. Open the chamber by an incision along the antero-ventral wall; probe the left atrio-ventricular passageway. Insert scissors into the *aortic trunk* and cut along its ventral wall (which is the medial side of the right atrium) into the left ventricle. Note the three *left semilunar valves* of the aortic trunk. Again one may have been damaged by the incision. Probe their thin cuplike walls. Note just above them two orifices. These are the exits of the *coronary arteries* to the musculature of the heart. Now examine the left ventricle. Observe the *trabeculae carne,* ridges of musculature, and the two *bicuspid* or *mitral valves* of the left atrio-ventricular canal. Close your eyes (they are tired); can you "see with the mind's eye" the organization of this remarkable double pump?

Before leaving the thorax, return to the preparation to observe the branching of the pulmonary trunk into *ductus arteriosus* and *pulmonary arteries.* Slit open the trunk and ductus. Is there a difference in size of the strand of latex in the ductus compared with that in a pulmonary artery? Explain. Deflect the left lung completely to the pig's right to observe the course of the *aorta.* Are you surprised at its size? Note the large vessel paralleling the aorta to the left and crossing ventral to the aorta as it arches to the right. This is the *hemiazygos vein.* Its entrance into the base of the superior vena cava was probably overlooked in removing the heart. What is the function of this vessel? Its origin? How does it differ from the hemiazygos in man? What is the explanation for differences in circulatory patterns?

An interesting dissection, if time permits, is the disclosure of the left *recurrent laryngeal nerve,* a branch of the vagus which arches under the ductus arteriosus and extends forward to the larynx. Explain this devious course of the nerve. What is meant by nerve towing? What is meant by differential rates of growth? Is there a difference between right and left sides? Explain.

4. ABDOMEN: Sever the *umbilical vein* a short distance anterior to the umbilicus. Make a rapid survey of the *liver, stomach, intestines,* and *spleen.* Identify the *central* and *lateral lobes* on both the right and left sides of the liver. Lift the right

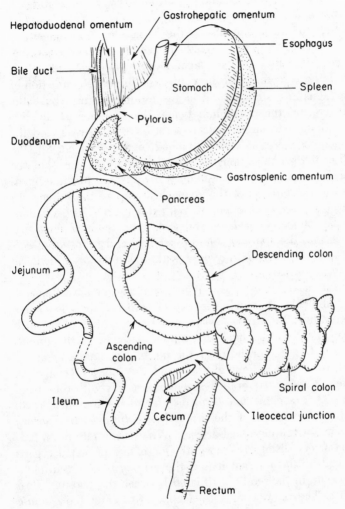

Fig. 74. Digestive tract (in part) of fetal pig.

central lobe to observe the *gall bladder* on its dorsal surface. Note the following omenta: *gastrohepatic, hepatoduodenal, gastrosplenic,* and *splenorenal.* With a probe free the *bile duct* running to the *duodenum* in the hepatoduodenal omentum. Identify the *pancreas,* an irregular, pebbly-surfaced organ between duodenum and stomach. Deflect stomach and spleen anteriorly to see more of the pancreas. Observe the junction of esophagus and stomach. Sever the esophagus just above the stomach and the duodenum below the entrance of the bile duct. Remove stomach, spleen, most of the pancreas, and the segment of the duodenum attached to the stomach. Slit the wall of the stomach along the *greater curvature,* wash out the contents, and examine the *pylorus,* the opening of the esophagus and the *mucosa* of the stomach. Compare the latter with that of a piece of duodenum. With the aid of figure 74 identify the connection of *ileum* and *colon* (the *ileocecal junction*), the *cecum,* the *spiraled colon, ascending* and *descending colons,* and the *mesentery proper* with the *superior mesenteric artery.*

ARTERIES AND VEINS POSTERIOR TO THE DIAPHRAGM. Relocate the umbilical vein which was severed earlier. Follow its course into the liver by tearing away the substance of the liver with forceps. Within the liver the venous pathway is the *ductus venosus.* Using a probe clear the peritoneum and connective tissue from the abdominal aorta and the inferior vena cava. Make out the major branches of these vessels with the aid of figure 73. Trace the *umbilical arteries* to the umbilical cord. Note the departure of the *internal iliac (hypogastric) artery* from one of the umbilical arteries. What is the fate of an umbilical artery distal to this point? One of the principal organs supplied by the internal iliac is the *urinary bladder,* an elongate organ in the fetal pig situated between the two umbilical arteries. The inferior vena cava appears to enter the *caudate lobe* (right side) of the liver. Pursue it by removing the substance of the liver. Explain this relationship embryologically, likewise the invariable position of the *inferior vena cava* to the right of the aorta, and the greater length of the left *renal vein* than the right one. Is there any other difference in the symmetry of the two renal veins? Find the stumps of *coeliac, superior mesenteric* and *inferior mesenteric arteries*. Which sections of the gut are supplied by these vessels?

UROGENITAL SYSTEM. Follow the *ureters* to the *urinary bladder,* previously identified. Sever a ureter below a kidney and remove the kidney, noting its peritoneal covering. What is meant by retroperitoneal? Make a frontal section of the kidney ventral to the attachments of the renal blood vessels, (see right kidney in figure 73). Observe the distribution of the branches of renal artery and vein. Open the *pelvis* of the kidney by inserting scissors into the ureter and cutting another frontal section through ureter and kidney. Trace the tip of the urinary bladder into the *allantoic stalk (urachus).* Remove a section of the pelvic girdle in the midline and the attached musculature, and spread the hind limbs apart to see the emergence of the *urethra* from the bladder. Trace the urethra to the tip of the *penis* in the male or the *urogenital sinus* of the female.

In the male identify a *scrotal sac* within which will be found a *testis, epididymis (head* and *tail), ductus deferens* emerging from the tail of the epididymis, and the *gubernaculum,* a ligament from the testis and tail of the epididymis to the scrotal wall. Observe the passage of the ductus deferens and spermatic blood vessels through the *inguinal canal* into the peritoneal cavity. Note the relationship of a ductus deferens to a ureter. Explain the looping of the ductus over (anteriorly) the ureter. Observe the entrance of the ducti deferentia into the urethra. The two pebbly organs at that point are the *seminal vesicles.* The prostate gland is not well developed at this time. The *bulbourethral glands,* however may be found alongside the pelvic part of the urethra.

In the female examine the opening of the *urogenital sinus* beneath the *genital papilla.* Trace the urogenital tract anteriorly identifying the point of attachment of the *urethra,* the *vagina* (between the entrance of the *urethra*) and the narrow-

231

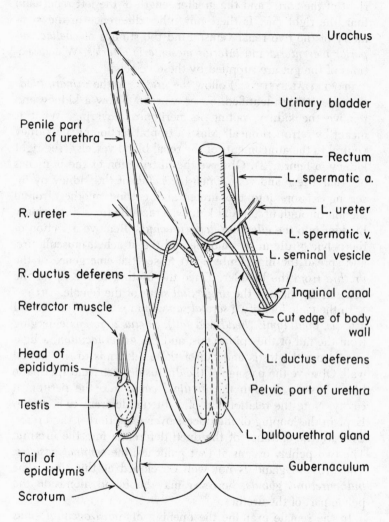

Fig. 75. Male urogenital system (in part) of fetal pig.

Labels (clockwise from top):
Urachus
Urinary bladder
Rectum
L. spermatic a.
L. ureter
L. spermatic v.
L. seminal vesicle
Inquinal canal
Cut edge of body wall
L. ductus deferens
Pelvic part of urethra
L. bulbourethral gland
Gubernaculum
Scrotum
Tail of epididymis
Testis
Head of epididymis
Retractor muscle
R. ductus deferens
R. ureter
Penile part of urethra

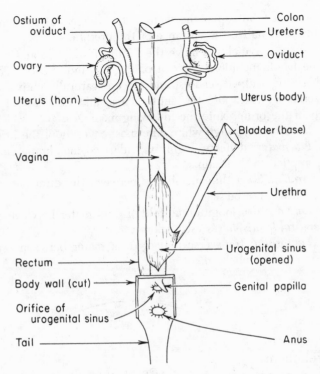

Ostium of oviduct

Ovary

Uterus (horn)

Vagina

Rectum

Body wall (cut)

Orifice of urogenital sinus

Tail

Colon

Ureters

Oviduct

Uterus (body)

Bladder (base)

Urethra

Urogenital sinus (opened)

Genital papilla

Anus

Fig. 76. Female urogenital system (in part) of fetal pig.

ing of the reproductive tract which is the boundary between vagina and *uterus (cervix)*. The section of the latter organ between the cervix and the point of bifurcation is known as the *body* of the uterus. The right and left arms are the *uterine horns* which extend laterad to the near vicinity of the *ovaries*. Locate the transition between a uterine tube and the corresponding very narrow *oviduct*. Trace the oviduct to its tip, a small bulb-like terminus the opening of which into the body-cavity is the *ostium* of the oviduct. What is the embryological origin of the patency of the oviduct? Name an analogous opening (albeit a temporary one) in another system of the embryo. (Hint:

233

neural tube) Slit open the wall of the urogenital sinus and vagina to visualize the entrance of the urethra.

5. BRAIN: Quickly remove the skin from the top and sides of the head; include the pinna of the ear and eyelids. Then with strong scalpel, scissors, and forceps carefully chip away the cranium to expose the entire dorsal and dorsolateral aspects of the brain. Note the following meninges.

Dura mater: tough white outermost covering of the brain.

Pia mater: delicate, highly vascularized membrane closely applied to the brain.

Arachnoid: a thin membrane between the dura and pia which will not be seen.

Falx cerebri: longitudinal fold of dura mater between the *cerebral hemispheres* (see below).

Tentorium: transverse fold of dura mater between cerebrum and cerebellum.

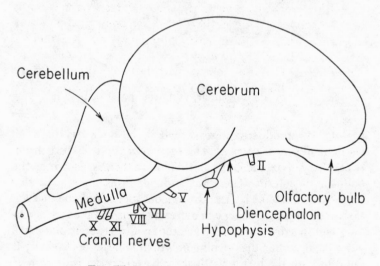

Fig. 77. Lateral view of brain of fetal pig.

Remove the meninges and identify the following parts of the brain. Use a smooth probe to separate the major divisions.

Cerebrum:

Cerebral hemispheres: a pair of large lobes, right and left.

Gyri and *sulci:* convolutions and fissures on the surface of the cerebral hemispheres (also on the cerebellum). Note that the pia mater extends into the sulci.

Olfactory bulbs: paired bodies situated at the anteroventral tips of the cerebral hemispheres. Careful dissection of the nasal area will disclose them. The *olfactory nerves* (bundle of fibers on each side) lead into olfactory bulbs through the ethmoid bone. Difficult to observe.

Corpus callosum: a commisure or fiber tract between the two cerebral hemispheres, seen by gently parting the two halves of the cerebrum. Only mammals among the vertebrates possess this structure.

Cerebellum: the second largest division of the brain, a pear-shaped body, flattened ventrally where it rests on the brain stem. Additionally there is a fiber tract, the *pons,* on the ventral side of the brain. It is not easily observed even after the removal of the brain from the cranium.

Brain stem. Carefully spread apart the cerebral hemispheres and the cerebellum so that you can inspect the *midbrain,* the middle segment of the brain stem. In dorsal aspect it consists of four lobes, the *colliculi.* The anterior pair are the superior colliculi, the posterior pair the inferior colliculi. Action of a probe will be needed to separate a superior colliculus from its associated inferior colliculus. What are their functions in the adult?

Gently free the brain along one side so that it can be lifted slightly to afford a ventrolateral view of the brain stem under the cerebrum and cerebellum. With a probe separate the cerebellum and *medulla,* the posterior part of the stem. In freeing the brain on one side you might have observed the *internal carotid artery* and the large *optic* and *trigeminal nerves.* Roll the brain toward the unattached side to note the *hypophysis* or pituitary gland, a small body at the tip of a short stalk, the *infundibulum.*

Now remove the brain from the cranial floor noting cranial nerves and internal carotid artery on the far side. Examine the ventral aspect of the brain. The *diencephalon,* the anteriormost division of the brain stem, can be identified as a vague ovoid region around and anterior to the hypophysis. Note the paired *anterior cerebral arteries* extending forward near the midline of the cerebrum, the unpaired *basilar artery* posterior to the hypophysis, and the *Circle of Willis.*

Cranial nerves. Some of the stumps of the cranial nerves may be observable. Usually the fetal brain is poorly fixed, however, so that delicate structures are not well preserved.

MATERNAL AND FETAL HEMOGLOBINS

Hemoglobin is globin, a globular protein, bound to four heme molecules. The globin in maternal hemoglobin (Hb-A) has four chains (2α and 2β) of amino acid residues. Each heme (one associated with each polypeptide chain) consists of a porphyrin ring with one iron atom (Fe II) in the middle. Fetal hemoglobin (Hb-F) differs from Hb-A: there are two γ polypeptide chains instead of two β chains (γ differs from β in the sequence of amino acids). The hemes of Hb-A and Hb-F are the same.

Oxygen bonds loosely with the iron to form bright red oxyhemoglobin. The ferrous atom is not oxidized to the ferric state. The degree of dissociation of oxyhemoglobin depends primarily on the concentration of oxygen (expressed as O_2 tension, or partial pressure, in mm Hg). At any given O_2 tension and pH of the blood Hb-F is more highly oxygenated than Hb-A. This *can be* an advantage to the fetus. Assume that the tension of oxygen in the placenta is about 40 mm Hg. Fetal blood *could* become 80% saturated with oxygen whereas at that tension maternal blood would be only about 60% saturated. Actual measurements show that fetal blood in the umbilical vein is probably not more than 60% saturated. The difference in "carrying capacity" of the two hemoglobins appears not to be related to differences in globins, but to factors in the plasmas and red blood corpuscles. In the human fetus less than 36 weeks of age 90% of the hemoglobin is the fetal type; thereafter, extending into postnatal life, Hb-F decreases about 4% per week. Even adult blood contains about 1% Hb-F.

236

The three most dramatic moments in the life cycle of a placental mammal are fertilization, birth, and death—events marked by sudden structural and physiological changes. Of those occuring at birth the most remarkable are the alterations in the circulatory system which Bradley Patten has characterized as a "perfect preparedness for birth built into the very architecture of the circulatory system during its development. Crowded into a few crucial moments is the change from water living to air living that in phylogeny must have been spread over eons of transitional existence." The lungs become inflated and functional and as a consequence profound shifts in blood flow occur that lead in turn to morphological modifications in the heart and certain arteries and veins, some of which are listed below.

1. Closure of the ductus arteriosus, the persisting left sixth aortic arch which serves during fetal life to. shunt blood from the pulmonary trunk to the arch of the aorta. Were it not for this bypass blood pressure on the right side of the fetal heart would be inordinately high. At birth the inflation of the lungs precipitously lowers the peripheral resistance to blood flow through the pulmonary bed of arterioles and capillaries and on to the left atrium via the pulmonary veins. As a consequence less blood flows through the ductus. Sphincter-like action of its muscular wall, triggered by the rise in oxygen content of the blood, causes a functional closure that is not complete until several days after birth. Then the vessel is slowly converted into the ligamentum arteriosum.

2. Closure of the foramen ovale, the passageway through the interatrial septa which shunts blood in the fetal heart from right to left atria (see p. 199). The term foramen ovale is here used in a general sense and not in *sensu strictu* as the aperture in septum secundum. Flow from right to left in the fetal heart is owing to a) the greater blood pressure on the right side because of pulmonary resistance (see above), and b) the structural arrangements of septa primum and secundum, the lower part of the former functioning as a flap-valve to the left of and against the opening in the latter. Following the onset of functional activity of the lungs there is an equalization of pressure in the two atria (by a drop on the right and a rise on the left). Consequently the two septa are pressed together and the foramen ovale is functionally closed. Later the septa may become fused.

3. Closure of the ductus venosus, the channel through the liver carrying blood returning from the placenta via the left umbilical

vein to the inferior vena cava. With the severing of the umbilical cord and the constriction of a muscular sphincter at the entrance of the umbilical vein into the liver the ductus venosus is no longer supplied. It collapses and later becomes transformed into the ligamentum venosum, a fibrous strand in the liver substance. Incidentally, the cessation of the umbilical stream through the ductus venosus is partly responsible for the drop in blood pressure in the right atrium discussed above.

4. Ligamentum teres or round ligament of the liver forms by the fibrous involution of the umbilical vein between the liver and the umbilicus (the point of attachment of the umbilical cord to the ventral body wall, commonly called the navel).

5. Lateral umbilical ligaments arise by fibrous involution of the distal segments of the two umbilical arteries, that is, the parts of these vessels along the ventral body wall. The proximal segments in the pelvis remain functional as the hypogastric or internal iliac arteries. Closure of the umbilical arteries occurs at birth. Stretching of the umbilical cord causes their longitudinal muscles to contract. A middle umbilical ligament arises from the urachus.

6. Other changes in the circulatory system begun during fetal life proceed after birth until the adult condition is attained, such as, for example, increase in blood pressure, decline in number of immature blood cells in the circulation, and the transition from fetal to adult type of hemoglobin. In connection with the last named it may be mentioned that fetal hemoglobin, which differs from the adult form in only its globin part, has a greater oxygen-carrying capacity than that of adult hemoglobin. This means that at oxygen tensions prevailing in the placenta fetal blood can attain a higher percentage of oxygen saturation than the maternal blood on the other side of the barrier.

PLACENTAL TYPES

Placentae are classified as follows.

    A. Morphological types, based upon distribution of chorionic villi.
        1. Diffuse. Villi scattered more or less evenly over chorion. Examples: horse, pig, tapir (perissodactyls in general); also lemur.
        2. Cotyledonary. Villi clustered into rosettes (called cotyledons). Examples: cow, sheep, deer (cud-chewing ungulates in general).
        3. Zonary. Villi restricted to a girdle or band around the

chorion. Examples: dog, cat, seal, raccoon (carnivores in general).

4. Discoidal. Villi aggregated into plaques at one pole (sometimes two poles) of chorionic vesicle. Examples: mouse, monkey (rodents and primates in general); also moles, shrews, and bats.

B. Histological types, based upon the thickness of the placental barrier.

1. Epithelio-chorial. Maximal thickness; no loss of layers; chorion in contact with uterine epithelium. Examples: Artiodactyla, Perissodactyla, Cetacea (whales), Lemuroidea.

2. Endothelio-chorial. Some erosion of uterine endometrium; chorion in contact with maternal blood vessels. Examples: Carnivora, Bradypodidae (sloths), European mole.

3. Hemo-chorial. Complete erosion on uterine side; chorion (very thin often) bathed with maternal blood. Examples: Rodentia, Primates (except lemurs), Sirenia (manatees), most Chiroptera (bats), most Insectivora.

LENGTH OF GESTATION

Animal	Taxonomic Group	Gestation (days)
Opossum	Marsupial	13°
Shrew	Insectivore	16
Mouse	Rodent	20
Rabbit	Lagomorph	32
Bat	Chiropteran	60
Cat, dog	Carnivore	63 (6 weeks)
Pig	Artiodactyl	119 (17 weeks)
Sheep, goat	Artiodactyl	147 (21 weeks)
Monkey	Primate	168 (24 weeks)
Bear	Carnivore	215 (31 weeks)
Human, ape	Primate	260 (37 weeks)°°
Cow	Artiodactyl	280 (40 weeks)
Horse	Perissodactyl	336 (48 weeks)
Donkey	Perissodactyl	365 (52 weeks)
Rhinoceros	Perissodactyl	548 (18 months)
Elephant	Proboscidean	640 (21 months)

° Born prematurely.

°° 280 days after onset of the last mensis (range of 250–310 days); 266 days (38 weeks) from fertilization; 260 from implantation, when gestation technically begins.

VI

GENERAL REVIEW QUESTIONS

1. Name the derivatives of the following.

1. Amnio-cardiac vesicles	32. Ventral mesentery
2. Epimyocardium	33. Cranial intestinal limb
3. Sinus venosus	34. Caudal intestinal limb
4. Endocardial cushions	35. Laryngotracheal groove
5. Blood islands	36. Mesogaster
6. Hemoblasts	37. Cloaca
7. Sino-atrial valves	38. Somite
8. Posterior cardinals	39. Cranial sclerotome
9. Subcardinals	40. Caudal sclerotome
10. Supracardinals	41. Mesonephric duct
11. Third aortic arches	42. Müllerian duct
12. Fourth aortic arches	43. Phallic sinus
13. Sixth aortic arches	44. Pelvic sinus
14. Vitelline veins	45. Allantois
15. Vitelline arteries	46. Urogenital fold
16. Anterior cardinals	47. Metanephrogenic blastema
17. Common cardinals	48. Ureteric bud
18. First pharyngeal pouch	49. Cortex and medulla of
19. Second pharyngeal pouch	gonad
20. Third pharyngeal pouch	50. Labio-scrotal swellings
21. Fourth pharyngeal pouch	51. Auditory vesicle
22. Fifth pharyngeal pouch	52. Neural crests
23. First visceral furrow	53. Prosencephalon
24. First visceral arch	54. Rhombencephalon
25. First closing plate	55. Basal plate
26. Dental lamina	56. Alar plate
27. Dental papilla	57. Rathke's pouch
28. Dental sac	58. Optic cup
29. Cranial liver diverticulum	59. Median nasal process
30. Caudal liver diverticulum	60. Lateral nasal process
31. Dorsal mesentery	

240

2. Name the sources of the following.

1. Arch of the aorta
2. Right subclavian artery
3. Posterior vena cava
4. Internal jugular
5. Renal veins
6. Ligamentum arteriosum
7. Ligamentum venosum
8. Ligamentum teres
9. Lateral umbilical ligaments
10. Vertebral artery
11. Thyrocervical artery
12. Internal mammary artery
13. Epigastric arteries
14. Hypogastric artery
15. Hepatic portal
16. Hepatic veins
17. Coronary sinus
18. Superior vena cava
19. Tongue
20. Diaphragm
21. Blood corpuscles
22. Upper jaw
23. Round ligament
24. Gubernaculum
25. Atrial septum
26. Sternum
27. Chondrocranium
28. Adrenal glands
29. Spleen
30. Neurilemma
31. Epiglottis
32. Arytenoid swelling
33. Ear ossicles
34. Coronary ligament
35. Duct of Wirsung
36. Duct of Santorini
37. Thyroid
38. Layers of the gut
39. Tympanic membrane
40. Parotid gland
41. Parathyroids
42. Thymus
43. Palatine tonsil
44. Cheeks
45. Teeth
46. Palate
47. Islands of Langerhans
48. Mediastinum
49. Male urethra
50. Median umbilical ligament
51. Ureter
52. Broad ligament
53. Pituitary body
54. Ventricular septum
55. Cerebral cortex
56. Iliac artery
57. Common iliac artery
58. External carotid
59. Middle sacral artery
60. Vagina

3. Complete the following table.

Name of nerve	Number in series	Name of ganglion (if not present write none)	Sensory, motor, or mixed?	Region of CNS to which it connects	One structure which it innervates
					Semicircular canals
				Telencephalon	
			Motor*		
		Geniculate			
	6th cranial				
Phrenic					

* Disregard the small number of proprioceptive fibers carried by "purely" motor nerves.

4. List the structures appropriate to the spaces in the chart below.

Fate in Male	Source	Fate in Female
Urethra		
	Mesonephric duct	
		Labia minora
Gubernaculum		
	Genital cord (fused Müllerian ducts)	
		Ovarian follicles of newborn

5. A. The horizontal axis below represents time in days of a non-pregnant and a pregnant human cycle. The M at 28 days indicates the average time of onset of menstruation at the close of the nonpregnant cycle. Parturition, marking the close of gestation, is represented by the letter P at 280 days. In similar fashion (using letters and numbers of days after zero) locate on this axis the following events:

Ovulation (O)
Implantation (I)
Neurulation (N)
Formation of primitive streak (S)

Antrum formation (A)
Appearance of extraembryonic mesoderm (E)

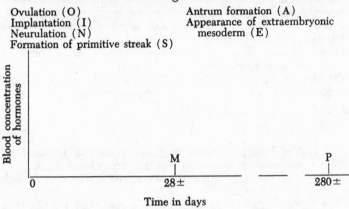

B. Now by means of variously colored curves show above the relative blood concentrations (vertical axis) of the following hormones in relation to time (horizontal axis):

Progesterone
Luteinizing hormone

Follicle stimulating hormone
Estrogen

6. Arrange the following items in proper anatomical order, as though encountered in a Caesarean section.

Syntrophoblast
Cytotrophoblast
Myometrium
Decidua capsularis
Ectoderm
Somatic mesoderm
Amniotic fluid

Stratum basale
Epidermis of the embryo
Extraembryonic coelom
Serosa (visceral peritoneum)
Decidua parietalis
Mesoderm of chorion

7. Rearrange the following items in proper time sequence. Visualize descent of the germ line in the female, the maturing

243

of one germ cell, its fertilization, early embryonic development, and associated events.

Oögonium
Primordial germ cell
Crossing over
Sperm entrance into egg
Mitotic divisions in germ line
Reduction to haploid number of chromosomes in female germ cell
Formation of first polar body
Ovulation
Formation of primary villi
Establishment of embryonic mesoderm
Fusion of decidua capsularis
and decidua parietalis
Development of allantois
Neurulation
Delamination of entoderm
Fusion of amnion and chorion
Origin of extraembryonic coelom
Implantation
Fall in blood concentration of progesterone
Rise to high blood concentration of FSH
Rise to high blood concentration of progesterone

8. Very briefly, *how* could one demonstrate that:

1. The lactogenic hormone (luteotrophin) is required for maintenance and activity of the corpus luteum.
2. Androgen is required by spermatozoa as they pass through the epididymis.
3. The cortex of the gonadal ridge contains a morphogenic factor which promotes ovarian differentiation.
4. The mesonephros of the mammalian fetus actually functions as an excretory organ.
5. Fetal and maternal hemoglobins differ in oxygen-carrying capacity.
6. The anlage (presumptive region) of the eye is irreversibly determined in the early neurula.
7. Placental permeability varies with the thickness of the placental barrier.
8. Specific interacting substances (gamones) are produced by gametes.
9. Primordial germ cells arise outside the gonad.
10. Ureteric bud is an organizer for the differentiation of the metanephric nephrotome.
11. A sea urchin egg (not yet cleaved) has been fertilized. (Give both a physiological and a morphological demonstration.)
12. Most of the blood entering the heart from the inferior vena cava passes through the foramen ovale.

13. The cleavage spindle lies lengthwise along the longest axis of the cytoplasmic mass of a blastomere.
14. Nuclei of the early blastomeres have no morphogenic significance (i.e., that they are qualitatively alike for development).
15. Prospective medullary plate in the beginning gastrula is undetermined for its future fate.
16. Body temperatures in most mammals prevent spermatogenesis.
17. Placental permeability increases with gestation.
18. Formation of neurilemma and myelin sheath require the presence of sheath cells.
19. Nerve fibers are outgrowths of neuroblasts and exhibit contact guidance.
20. The sex of a toad and of a pullet can be reversed.

9. Draw and label fully:

1. The fetal membranes of the chick embryo, show germ layers by color (blue: ectoderm; red: mesoderm; yellow: entoderm).
2. A *cross* section of the human uterus containing a two-month embryo. Show all layers of the uterus, with a part of the endometrium in detail; deciduae; fetal membranes; extraembryonic coelom and other cavities.
3. A cross section of the umbilical cord at a level near the embryo.
4. The heart (diagrammatically) and the arteries and veins anterior to the diaphragm of the near-term fetal pig.
5. A schematized frontal view of the buccal and pharyngeal regions of the embryonic gut showing the primordia of the pharyngeal derivatives (e.g., thymus) and of tongue and epiglottis.
6. A diagrammatic sketch (ventral view) of the six pairs of aortic arches in the mammalian embryo connected medially to the ventral aorta (aortae) and laterally to the dorsal aortae. Indicate those segments of the above vessels which persist.
7. An outline sketch of the vertebrate brain (either straight or with cranial flexure) as seen from the side. Label the five major divisions of the brain and their cavities. In addition label the following by suggested symbols:

Isthmus (Is)　　　　　　Olfactory bulbs (O)
Velum transversum (V)　Superior colliculi (SC)

Infundibulum (In)	Auditory vesicle (AV)
Hypophysis (H)	Gasserian ganglion (Ga)
Epiphysis (Ep)	Geniculate ganglion (Ge)
Pons (P)	Lamina terminalis (L)

Show also the stumps of the following cranial nerves, labeled with appropriate Roman numerals.

Trochlear	Trigeminal
Glossopharyngeal	Oculomotor
Optic	Olfactory
Facial	

8. A cross section of the spinal cord with spinal nerve shown on one side including one of each of the following types of neurones: somatic afferent, somatic efferent, visceral afferent, and visceral efferent. Label regions of gray and white matter and the branches of the spinal nerve.

9. A ventral view of the digestive system and the dorsal mesentery. Differentiate between regions of the mesentery which become fused to the body wall and those which remain as mesenteries or omenta.

10. Lateral view of the amphibian fate map. The avian fate map as seen from above.

11. A cross section of the embryo at the level of the stomach showing the dorsal and ventral mesenteries and associated viscera.

10. Name in order the specific parts of the circulatory system of the 48-hour chick embryo:

> a. Carrying blood from the ventral aorta, directly through the visceral arch between 2nd and 3rd pharyngeal pouches, and on to the walls of the diencephalon *with oxygenation en route*.
>
> b. Traversed by a blood corpuscle after breaking loose from a blood island and passing via the mandibular arch to the spinal cord at some posterior level (e.g., level of the 20th somite) and thence to the sinus venosus.

11. List in order the vessels and chambers of the heart of the *most probable* route for blood flowing in the human fetus:

> a. From left adrenal gland to the right half of the lower jaw.
> b. From the left thoracic wall to the cerebellum.

12. The "purest" blood in the fetal circulation is that en

route to the fetus from the placenta. By the time this stream reaches the lower dorsal aorta it has been "diluted" at several points by incoming streams of "less pure" blood. List, in order, the places in the fetal circulation at which dilution occurs and for each place name the diluting stream.

13. List the following parts of the fetal circulatory system in order with respect to relative oxygen content of the blood: innominate artery, ductus venosus, dorsal aorta (abdominal part), umbilical vein, right ventricle, left ventricle, left innominate vein.

14. Diagram the plan of fetal circulation.

15. List the structural and physiological changes in the fetus at birth or shortly thereafter.

16. Of what value is embryology in determining phylogenic relationships and the course of evolution of the vertebrates?

17. Name organisms to fit the key below.

   I. Amnion present; 12 cranial nerves develop; caudal region only of holonephros persists as adult excretory organ
     A. Miolecithal egg; chief nitrogenous waste of embryo is urea
        1. Allantois vestigial; maternal blood bathes chorionic villi
           a. Müllerian ducts fuse to form a unitary uterus (simplex); left common cardinal drops out (largely if not entirely) .... (1)
           b. Each Müllerian duct forms a uterine tube (bipartite); left common cardinal persists as left superior vena cava .... (2)
        2. Allantois large; no erosion of maternal vessels by trophoblast
           a. Chorionic villi arranged in clusters (cotyledons) .... (3)
           b. Chorionic villi arranged in an equatorial band .... (4)
     B. Megalecithal egg; chief nitrogenous waste of embryo is or presumed to be uric acid

1. Young nourished by milk; muscle buds invade transverse septum .... (5)
2. Young not milk-fed; transverse septum remains non-muscular
   a. Right fourth aortic arch contributes to the adult aorta (systemic arch); optimal temperature for development is around 100° F. (6)
   b. .Left fourth aortic arch contributes to an adult aorta (systemic arch); development would not occur at a temperature of 100° F. .... (7)

II. Amnion absent; hypoglossal and spinal accessory nerves do not develop; no true kindey (metanephros) in adult
   A. Early cleavages meroblastic (incomplete) .... (8)
   B. Early cleavages holoblastic (complete)
      1. Regulative egg (separation of first two blastomeres usually gives equal and normal development)
         a. Involution the dominant gastrular process; schizocoelous (coelom formation by splitting of mesoderm) .... (9)
         b. Invagination the dominant gastrular process: enterocoelous (coelom formation by archenteric outpocketings) ....(10)
      2. Mosaic egg (separation of first two blastomeres gives unequal and defective development)
         a. Protostomia (blastopore becomes mouth); spiral cleavage; notochord absent ....(11)
         b. Deuterostomia (mouth formed from stomodeum); radial cleavage; notochord forms from mesoderm ....(12)

18. Construct a key to distinguish: opossum, duckbill (monotreme), dog, rat, gorilla, cow, salamander, sea urchin, amphioxus, sparrow, lizard.

19. Each of the following sets of embryological features characterize a group of animals with the exception of one false item in each set. Name the inconsistent items and the groups of animals which are characterized after the irregular items are excluded.

A. Cleidoic egg, meroblastic cleavage, entoderm formation by delamination, megalecithal egg, urea the principal nitrogenous waste, third cleavage plane vertical.

B. Amnion formed by cavitation, holoblastic cleavage, third cleavage plane horizontal, regulative egg, gastrulation principally by invagination, miolecithal.

C. Anamnia (no amnion), protostomia (mouth derived from blastopore), medialecithal, indeterminate cleavage, sperm enters secondary oocyte, spinal accessory and hypoglossal nerves do not develop.

D. Non-deciduate, chorionic villi arranged in cotyledons, endotheliochorial (or syndesmochorial) placenta, allantois vestigial, fetal blood sugar principally fructose.

20. Give an embryological explanation for each of the following.

1. The omental bursa communicates with the peritoneal cavity on the right side only.
2. The cement of the teeth is firmly attached to the dentary or maxillary bone.
3. The transverse colon lies anterior and ventral to the jejunum and ileum.
4. The liver is fused to the diaphragm.
5. Parietal pleura and parietal pericardium are fused together.
6. Both right and left common carotids arise from the brachiocephalic (innominate) artery in the pig. Speculate!
7. There are two pleural cavities but only one pericardial and one peritoneal cavity.
8. The eustachian tubes, although derived from the first pharyngeal pouches, open into the respiratory passage instead of into the digestive tract.
9. The blood from the thoracic body wall returns to the heart via the superior vena cava instead of via the inferior vena cava.
10. The pulmonary trunk, although carrying blood from the right ventricle, lies for the most part to the left of the aortic trunk.
11. The left renal vein is longer than the right one.
12. The omental bursa is bounded in part by the transverse mesocolon.
13. The superior mesenteric artery supplies the whole of the

249

small intestine below the duodenum and the ascending colon of the large intestine.

14. The internal mammary artery is a branch of the subclavian artery.
15. Nerve fibers to the masseter muscle are regarded as visceral motor.
16. The ureter, although developmentally an outgrowth from the nephric duct, discharges directly into the urinary bladder.
17. Identical twins have two amnions but only one chorion.
18. Orchidectomy leads to sex reversal in the toad but not in the bird.
19. The phrenic nerves are quite long spinal nerves.
20. Myosepta are attached to the middle of vertebrae instead of to their ends,
21. The auditory nerve has distinct ganglia but not the olfactory and optic nerves.
22. The gonad of the female fetus at term is more advanced in development than that of the newborn male.
23. The pancreatic duct (Wirsung) opens into the bile duct rather than into the duodenum directly.
24. Some mammals (e.g. man) have one superior vena cava whereas others (e.g. rat) have two.
25. The tunica albuginea is the outermost covering of the testis but not of the ovary.
26. The right lung has a middle lobe but not the left lung.
27. The left recurrent laryngeal nerve of the vagus is longer than the right one.
28. The diaphram, although situated at the level of the first lumbar segment, is innervated by nerves from cervical segments 3-5.
29. Facial muscles are innervated by the seventh cranial nerve.
30. The epithelium of the tongue is innervated by four cranial nerves (V, VII, IX, and X) whereas the musculature of the tongue is supplied by one (XII).
31. The splenorenal ligament is attached to the left kidney.
32. Apposition of enamel and dentine without intervening cellular elements.
33. The transverse colon and stomach are connected by an omentum (gastro-colic ligament).
34. The ductus deferens loops over the ureter before entering the urethra.

35. The vertebral, thyrocervical axis, and internal mammary arteries, although supplying many segments of the body, are branches of the subclavian artery instead of branches of the dorsal aorta.

36. In some mammals (e.g. primates) the excretory and reproductive systems of the female open separately to the exterior; in others (e.g. carnivores) they have a common orifice; in still others (monotremes) digestive, excretory, and reproductive systems terminate together.

37. Some mammals (rodents) exhibit a subdivided uterus (bipartite); others (ungulates) a bilobed uterus (bicornuate); still others (primates) a single dome-shaped organ (simplex).

38. The suspension of incus, malleus, and stapes within the middle ear.

39. Rods and cones are situated on the inner (medial) surface of the retina whereas sensory cells are usually on the outer (distal) surface of an epithelial layer.

40. Accessory adrenal-like chromaffin bodies (paraganglia) are scattered along the chain of sympathetic ganglia.

41. The transverse colon is freely movable but not the ascending and descending colons.

42. Enamel does not cover the roots of teeth; neither, on the other hand, is cement found in the crown of the tooth.

43. Organs such as heart, lungs, liver, and intestines do not really lie in the coelom, being covered with pericardium, pleura, or peritoneum.

44. The superior parathyroids are derived from the fourth pharyngeal pouches whereas the inferior parathyroids, which lie caudal to the first pair, are derived from more anterior pouches, namely, the third pouches.

45. Nerve fibers in the gray matter of the brain are gray, whereas those in the white matter and in the peripheral nervous system (excepting postganglionic fibers of the autonomic system) are white.

46. Hypospadius (a developmental anomaly in the male in which the urethra is open via clefts on the under surface of the penis).

47. Motor fibers to the diaphragm are classified as somatic motor.

48. The receptors of the vertebrate lateral eye project from the "backside" of the retina whereas those of the median third eye of reptiles extend into the cavity of the organ.

251

49. If the skin of a developing amphibian embryo is removed from three segments of the body the spinal ganglia supplying that area show an approximate 60% reduction in nerve cell bodies, whereas if the myotomes of the same segments be removed, the corresponding ganglia show about a 40% reduction in nerve cell bodies.

50. Unlike most muscles which are mesodermal in origin, the pupillary muscles of the eye are ectodermally derived.

51. Supernumerary mammary glands are most commonly found along a line from the inguinal (groin) to the axillary (armpit) regions.

52. Meckel's diverticulum of the ileum (a developmental anomaly consisting of a sacculation of the ileum above the ileocecal valve which may be a blind pouch or it may retain an opening at the umbilicus).

53. Sex reversal of the gonads sometimes occurs in the female member of a heterosexual pair of twins in cattle, but not always and rarely if at all in other mammals.

54. (This is purely hypothetical.) A salamander shows on one side the following abnormalities: albinism owing to the absence of pigment cells in the integument, insensitivity of the skin to touch, lack of chromaffin (adrenal) tissue.

55. The right suprarenal vein is a tributary of the inferior vena cava but the left suprarenal vein empties into the left renal vein.

56. The liver although arising from the duodenal region of the gut is connected to the stomach (by the gastrohepatic ligament).

57. The teeth which replace the milk molars are called premolars.

58. The optic nerve fibers originate from the vitreal surface of the retina instead of the tapetal surface, whereas olfactory nerve fibers lead directly from the back surface of the olfactory epithelium.

59. Cholecystectomy (surgical removal of the gall bladder) does not interfere with the flow of bile from liver to duodenum.

60. The cavities of outer, middle, and inner ears are not continuous.

61. The female urethra is homologous with only the proximal segment of the male urethra.

62. Hare lip and cleft palate.

63. The right pulmonary artery (a derivative of the right sixth aortic arch) receives blood from the pulmonary trunk which, at the point of its branching into the pulmonary arteries, is derived from the *left* side of the old ventral aorta or aortic sac.

64. Neuralcrestectomy in the region of the future 10th cranial nerve results in a reduction or absence of innervation of the region of the skin normally supplied by the vagus but the operation does not affect the number of vagal sensory fibers to the heart and gut.

65. Although the conus has four bulbar ridges only three semilunar valves form in the aortic trunk.

66. The hepatic portal vein has a spiraled course around the gut.

67. Blood may enter the basilar artery anteriorly or posteriorly.

68. Although the ureteric bud is a posterior outgrowth of the nephric duct, the adult derivative of the former terminates anterior to the adult derivative of the latter.

69. The transverse colon lies anterior and ventral to the jejunum and ileum.

70. The pancreatic duct opens into the bile duct in some mammals (sheep) whereas in others (pig) it enters the duodenum directly.

71. Dentine can be reformed but not enamel once the tooth has erupted.

72. Protrusion of the ileum through the diaphragm to lie against the left lung.

73. At an O_2 tension of 40 mm. Hg a sample of umbilical blood can become about 80 percent saturated with oxygen, whereas blood from the maternal uterine vein can become only about 70 percent saturated at the same partial pressure.

74. The third, fourth, and sixth cranial nerves have no ganglia.

75. The rare anomaly of bifid tongue in which there are two tips.

21. For the following exercises (A to D) name the "unknown" animal as specifically as you can from the five embryological facts given and then very briefly show how each fact enabled you to "narrow down" the alternatives by eliminating certain possibilities.

A. This animal exhibits:
   1. Miolecithal egg
   2. Radial cleavage
   3. Entoderm formation by delamination
   4. Amnion formation by cavitation
   5. Left common cardinal disappears or is very greatly reduced.

B. This animal exhibits:
   1. Coelom formation by splitting of mesoderm (schizocoel)
   2. Meroblastic cleavage
   3. Cleidoic egg (i.e., nitrogenous wastes retained within the developmental system as for example inside the shell)
   4. Development requires relatively high temperature (i.e., incubation)
   5. Ovariectomy of juvenile leads to sex reversal.

C. This animal exhibits:
   1. Hypoglossal and spinal accessory nerves arising from cranial neuroblasts
   2. Dental lamina develops different kinds of teeth (e.g., incisors, molars)
   3. Cotyledonary placenta
   4. Müllerian ducts disappear or remain as vestiges only; nephric (Wolffian) ducts differentiate
   5. Phallic sinus remains open permanently; sex chromosomes XX.

D. This animal exhibits:
   1. Deuterostomia (mouth formed from stomodeum, not from blastopore)
   2. Central nervous system formed by folding of medullary place, hence hollow
   3. Third cleavage plane normally horizontal
   4. Allantois and metanephros (true kidney) do not form.
   5. First and second aortic arches drop out; fourth arches form paired aortae.

22. (This is a variation of question 13). Assume that one could detect differences in oxygen content of blood in various parts of the circulatory system of the human fetus shortly before birth. Rearrange the items below in order of blood quality, that is, from high to low oxygen content. Bracket to-

gether those of equal quality and disregard vessels which have ceased to function.

Ductus venosus	Left external carotid
Left ventricle	Left posterior cardinal
Umbilical arteries	External iliac arteries
Left umbilical vein	Blood passing through foramen ovale
Right umbilical vein	Inferior vena cava at entrance into right atrium
Right internal carotid	Part of inferior vena cava derived from subcardinal

23. Defend the statements below if true or criticise them if they are false, partly true, or incomplete.

1. The concept of preformation is untenable.
2. Ontogeny is the result of an interaction between internal (hereditary) and external (environmental) factors. (Environmental as used here means any factor outside the developing embryo.)
3. Induction (i.e., the organizing influence of one district of an embryo upon another region) is the universal and dominant causation of differentiation.
4. If one were to disregard inductive influences and consider only causal factors of differentiation within a given cell, it might be said that the nucleus is more important than the cytoplasm.
5. Activation of the egg either by sperm entrance or by parthenogenesis involves important changes in the cortex of the egg.
6. The fate of an embryonic blood vessel is largely the consequence of changes in pressure of the blood flowing through the vessel.
7. The volume of oxygen transferred from mother to fetus in placental mammals depends upon not one but several factors.
8. The foramen ovale (interatrial connection as used here) is a remarkably important and effective "invention" of nature.
9. The mesonephros of a mammalian embryo is not functional as an excretory organ, all nitrogenous wastes being removed from the embryonic circulation by the placenta.
10. Through the study of regeneration of nerve fibers in an

255

adult mammal one can learn how a functional peripheral nervous system is established in the embryo.

11. It is an axiom of development that the fate of a cell is the function of its position in the system.

12. The mechanics of gastrulation (i.e., the types of cell movements involved in the establishment of entoderm and mesoderm as discrete layers) varies among animals because of a variation in the amount of yolk present.

13. Molecules with a molecular weight of less than 350 pass the placental barrier by diffusion.

14. The difference between mosaic and regulative eggs lies in the patterns of distribution of materials in the egg.

15. Differentiation, the crux of development, is the result of the inductive action of organizers.

16. The placental barrier, which is known to vary in thickness, acts like a correspondingly thick or thin collodion membrane.

17. Blood normally ceases to flow across the ductus arteriosus within minutes after birth.

18. The ureteric bud is an organizer for the metanephros.

19. An organizer (e.g. dorsal lip) must be in contact with the reacting system (e.g. competent ectoderm) to effect an induction.

20. The nuclei of the cells of an embryo (e.g. gastrula or neurula) are different so far as their developmental potencies are concerned, the implications of the mitotic process to the contrary notwithstanding.

BIBLIOGRAPHY

Abercrombie, M. and Jean Brachet. *Advances in Morphogenesis*. New York and London: Academic Press. A multi-volume open-ended series, beginning in 1962.

Arey, Leslie B. *Developmental Anatomy*. 7th ed. Philadelphia and London: W. B. Saunders, 1965.

Balinsky, B. I. *An Introduction to Embryology*. Philadelphia and London: W. B. Saunders, 1960.

Ballard, William W. *Comparative Anatomy and Embryology*. New York: The Ronald Press, 1964.

Barth, Lester G. *Embryology*. Rev. ed. New York: Dryden Press, 1953.

Bell, Eugene, ed. *Molecular and Cellular Aspects of Development*. New York, Evanston, and London: Harper and Row, 1965.

Brachet, Jean. *The Biochemistry of Development*. New York: Pergamon Press, 1960.

Davies, Jack. *Human Developmental Anatomy*. New York: The Ronald Press, 1963.

Dawes, Geoffrey S. *Foetal and Neonatal Physiology*. Chicago: Year Book Medical Publishers, 1968.

DeHaan, Robert L. and Heinrich Ursprung, eds. *Organogenesis*. New York: Holt, Rinehart and Winston, 1965.

Deuchar, Elizabeth M. *Biochemical Aspects of Amphibian Development*. New York: John Wiley and Sons, 1966.

Gilchrist, Francis G. *A Survey of Embryology*. New York: McGraw-Hill, 1968.

Hamilton, W. J., J. D. Boyd, and H. W. Mossman. *Human Embryology*. 3d ed. Baltimore, Md.: Williams and Wilkins, 1963.

Huettner, Alfred F. *Fundamentals of Comparative Embryology of the Vertebrates*. Rev. ed. New York: Macmillan, 1949.

Lillie, F. R. and Howard L. Hamilton. *Lillie's Development of the Chick: An Introduction to Embryology*. 3d. ed. New York: Henry Holt and Co., 1952.

257

Monroy, Alberto. *Chemistry and Physiology of Fertilization.* New York: Holt, Rinehart and Winston, 1965.

Nelson, Olin E. *Comparative Embryology of the Vertebrates.* New York: Blakiston, 1953.

Patten, Bradley M. *Early Embryology of the Chick.* 4th ed. New York: Blakiston, 1953.

Patten, Bradley M. *Embryology of the Pig.* 3d. ed. New York: Blakiston, 1953.

Patten, Bradley M. *Foundations of Embryology.* 2d. ed. New York: McGraw-Hill, 1964.

Patten, Bradley M. *Human Embryology.* 2d. ed. New York: Blakiston, 1953.

Rugh, Roberts. *The Frog.* New York: McGraw-Hill. 1953.

Saxén, Lauri and Sulo Toivonen. *Primary Embryonic Induction.* New York and London: Academic Press, 1962.

Spemann, Hans. *Embryonic Development and Induction.* London: Oxford University Press, 1938.

Thomas, James Blake. *Introduction to Hunman Embryology.* Philadelphia: Lea and Febiger, 1968.

Torrey, Theodore W. *Morphogenesis of the Vertebrates.* New York and London: John Wiley and Sons, 1962.

Waddington, C. H. *New Patterns in Genetics and Development.* New York and London: Columbia University Press, 1962.

Weber, Rudolf, ed. *The Biochemistry of Animal Development.* New York: Academic Press. Vol. I, Descriptive Biochemistry of Animal Development, 1965; Vol. II, Biochemical Control Mechanisms and Adaptations in Development, 1967.

Willier, Benjamin H., Paul Weiss, and Viktor Hamburger. *Analysis of Development.* Philadelphia and London: W. B. Saunders, 1955.

Willier, Benjamin H., and Jane M. Oppenheimer, eds. *Foundations of Experimental Embryology.* Englewood Cliffs, New Jersey: Prentice-Hall, 1964.

Witschi, Emil. *Development of Vertebrates.* Philadelphia and London: W. B. Saunders, 1956.

PAPERBACK EDITIONS

Austin, C. R. *Fertilization.* Englewood Cliffs, New Jersey: Prentice-Hall, 1965.

Barth, Lucena Jaeger. *Development: Selected Topics.* Reading, Massachusetts, Palo Alto, and London: Addison-Wesley, 1964.

Ebert, James D. *Interacting Systems in Development.* New York: Holt, Rinehart and Winston, 1965.

Flickinger, Reed A. *Developmental Biology*. Dubuque, Iowa: Wm. C. Brown, 1966.

Grobstein, Clifford. *The Strategy of Life*. San Francisco: W. H. Freeman and Company, 1965.

Whittaker, J. Richard. *Cellular Differentiation*. Belmont, California: Dickenson, 1968.

INDEX

The following symbols are used to identify different embryos: *s*, starfish; *tw*, tubeworm; *m*, mussel; *a*, early amphibian; *f*, 10 mm. frog; *24c*, 15-25 hour chick; *33c*, 33 hour chick; *48c*, 48 hour chick; *72c*, 72 hour chick; *10p*, 10mm. pig; *fp*, fetal pig.

upper: 10*p*, 189

lagena, *f*, 76
lamina terminalis, 72*c*, 155
laryngotracheal groove: 48*c*,
    122, 138; 72*c*, 157
larynx, *fp*, 223
lateral limiting sulcus, 10*p*, 198
lateral ventricle, 10*p*, 178
lens: *f*, 73; 48*c*, 121, 133; 72*c*,
    153; 10*p*, 180
lesser omentum, 10*p*, 211
limb bud: 72*c*, 171; 10*p*, 216
liver: *f*, 80; 48*c*, 140; 72*c*, 159;
    10*p*, 191; *fp*, 228-230, 237-
    238
lung, *fp*, 223
lung bud: f, 80; 48*c*, 138; 72*c*,
    159; 10*p*, 190
lutenizing hormone, 24-28
luteotrophin, 25

macromeres, 53
mandibular process: 72*c*, 157;
    10*p*, 186
mantle layer, brain: *f*, 68; 72*c*,
    156; 10*p*, 212
maxillary process: 72*c*, 157; 10*p*,
    186
mediastinum: 72*c*, 167; 10*p*, 209
medulla, *fp*, 235
meiosis, 15-16, 33
melanophores, *f*, 69
menstrual cycle, 24-27
mesencephalon: 67; *f*, 71; 33*c*,
    103, 110; 48*c*, 117, 129,
    130; 72*c*, 151; 10*p*, 177;
    fp, 235
mesenchyme: *s*, 41; *f*, 71; 24*c*,
    93, 95; 48*c*, 129
mesentery, 165, 209
  caval, 10*p*, 201, 211
  derivatives of, 165

dorsal: 48*c*, 144; 72*c*, 166;
    10*p*, 205, 208, 210, 212
ventral, 72*c*, 166; 10*p*, 210,
    211
mesentery proper, 10*p*, 212; *fp*,
    230
mesocardium: 33*c*, 112; 48*c*,
    135; 72*c*, 166
mesocolon, 10*p*, 212
mesoduodenum, 10*p*, 212
mesogaster (greater omentum),
    10*p*, 211
mesonephric duct (Wolffian
    duct): *f*, 81; 48*c*, 141; 72*c*,
    167; 10*p*, 204
mesonephros: 48*c*, 141; 72*c*,
    167; 10*p*, 201
metanephros, 10*p*, 202, 205
metencephalon: 67; *f*, 77; 33*c*,
    103; 48*c*, 120, 130; 72*c*,
    147; 10*p*, 177
micromeres, 53
midbrain, *See* mesencephalon
midgut: 24*c*, 99; 48*c*, 140
mitral valves: 10*p*, 198
myelencephalon: 67; *f*, 74; 33*c*,
    103; 48*c*, 120, 129; 72*c*,
    147; 10*p*, 176
myotome: *f*, 78; 48*c*, 123; 72*c*,
    171; 10*p*, 214

nasal organs: *f*, 69; 10*p*, 181
nasal pit: 72*c*, 155; 10*p*, 181
nasal placode, 48*c*, 135
nasal process, 10*p*, 181
nephric system, general, 142-143
nephrostome: *f*, 80; 48*c*, 142,
    143; 72*c*, 168
nephrotome: *a*, 60; 24*c*, 99; 48*c*,
    141, 144
nerves
  abducens (VI), 10*p*, 184
  auditory (VIII): *f*, 76; 10*p*,